Parametric Optimization:
Singularities,
Pathfollowing and Jumps

PARAMETRIC OPTIMIZATION: SINGULARITIES, PATHFOLLOWING AND JUMPS

J. Guddat

and

F. Guerra Vazquez,

with

H. Th. Jongen

Springer Fachmedien Wiesbaden GmbH

CIP-Titelaufnahme der Deutschen Bibliothek

Parametric optimization: singularities, pathfollowing, and
jumps / J. Guddat and F. Guerra Vazquez with H. Th. Jongen.

ISBN 978-3-519-02112-4   ISBN 978-3-663-12160-2 (eBook)
DOI 10.1007/978-3-663-12160-2

NE: Guddat, Jürgen [Mitverf.]; Guerra Vazquez, Francisco [Mitverf.];
Jongen, Hubertus Th. [Mitverf.]

Library of Congress Cataloging-in-Publication Data:

Guddat, Jürgen.
      Parametric optimization : singularities, pathfollowing, and jumps
   / by J. Guddat, F. Guerra Vazquez, H. Th. Jongen.
          p.   cm.
      Includes bibliographical references (p.   ) and index.
      ISBN 978-3-519-02112-4
      1. Mathematical optimization.   2. Homotopy theory.   I. Guerra
   Vazquez, F.   II. Jongen, H. Th. (Hubertus Th.),   1947–
   III. Title.
   QA402.5.G83   1990   90-12491
   519.3—dc20   CIP

British Library Cataloguing in Publication Data:
Guddat, J.
   Parametric optimization : singularities, pathfollowing and jumps.
   1. Mathematics. Optimization. Algorithms
   I. Title   II. Guerra, Vazquez, F.   III. Jongen, H. Th.
   519.3

      ISBN 978-3-519-02112-4

Typeset by Thomson Press (India) Ltd, New Delhi

Contents

Preface   vii

1 Introduction   **1**
1.1 A preliminary survey on solution algorithms in one-parametric optimization   1
1.2 Some motivations   10
1.3 Summaries of Chapters 2–6   18

2 Theoretical Background (by H. Th. Jongen)   **20**
2.1 Preliminary outline   20
2.2 Unconstrained optimization problems   21
2.3 Constraint sets   28
2.4 Critical points, stationary points, stability   31
2.5 Generic singularities in one-parametric optimization problems   41
2.6 The approach via piecewise differentiability   53

3 Pathfollowing of Curves of Local Minimizers   **56**
3.1 Preliminary outline   56
3.2 The estimation of the radius of convergence   59
3.3 An active index set strategy   68
3.4 The algorithm PATH I and numerical results   83

4 Pathfollowing Along a Connected Component in the Karush–Kuhn–Tucker Set and in the Critical Set   **91**
4.1 Preliminary outline   91
4.2 Pathfollowing in the Karush–Kuhn–Tucker set   92
4.3 The algorithm PATH II and numerical results   96
4.4 Pathfollowing in the critical set   102
4.5 The algorithm PATH III   111

5 Pathfollowing with Jumps in the Set of Local Minimizers and in the Set of Generalized Critical Points   **115**
5.1 Preliminary outline   115
5.2 Jumps in the set of local minimizers and the algorithm JUMP I   117
5.3 Jumps in the critical set and the algorithm JUMP II   135

6 Applications **147**
6.1 Preliminary outline 147
6.2 On globally convergent algorithms 148
6.3 On global optimization 160
6.4 On multi-objective optimization 166

References and Further Reading 175
Glossary of Symbols and Some Assumptions 187
Index 190

Preface

This volume is intended for readers who, whether they be mathematicians, workers in other fields or students, are familiar with the basic approaches and methods of mathematical optimization. The subject matter is concerned with optimization problems in which some or all of the individual data involved depend on one parameter. Such problems are called one-parametric optimization problems. Solution algorithms for such problems are of interest for several reasons. We consider here mainly applications of solution algorithms for one-parametric optimization problems in the following fields:

(i) globally convergent algorithms for nonlinear, in particular non-convex, optimization problems,

(ii) global optimization,

(iii) multi-objective optimization.

The main tool for a solution algorithm for a one-parametric optimization problem will be the so-called pathfollowing methods (also called continuation or homotopy methods) (cf. Chapters 3 and 4). Classical methods in the set of stationary points will be extended to the set of all generalized critical points. This could be helpful since the path of stationary points stops in this set, but there is a continuation in the broader set of generalized critical points. However, it will be shown that pathfollowing methods only are not successful in every case. This is the reason why we propose to jump from one connected component in the set of local minimizers and generalized critical points, respectively, to another one (Chapter 5).

For both pathfollowing methods and jumps we need information on the structure of the set (depending on one parameter) of all local minimizers, stationary points and generalized critical points, respectively. Furthermore singularities (degeneracies) play an important role as the theoretical basis for pathfollowing methods with jumps. Chapter 2 contains the theoretical background. Two generic classes (the class of Jongen–Jonker–Twilt and the class of Kojima–Hirabayashi) will be introduced there. For the first class, all the various cases of singularities are known. Information on the singularities could be useful to construct a jump to another connected component. If we have jumps in all cases, then the fundamental problems with respect to (i), (ii) and (iii) will be solved for the considered class. From this point of view it is no surprise that we do not have proposals for jumps in any case.

The content of this book is mainly based on long common research of (i) H. Gfrerer, J. Guddat, Hj. Wacker and W. Zulehner (Kepler University, Linz, and Humboldt University, Berlin) with respect to pathfollowing methods in the sets of local minimizers and stationary points, (ii) H. Th. Jongen, P. Jonker and F. Twilt (TH Twente Enschede) on the theoretical foundation (e.g. singularity theory), and (iii) the three authors together and also with D. Nowack (Humboldt University, Berlin) and J. Rückmann (TH Leipzig), and results given by R. Lehmann and J. Rückmann in their doctoral theses at the Humboldt University, Berlin, and the TH Leipzig, respectively. Moreover, the results of M. Kojima and P. Hirabayashi (Tokyo) play an important role. Chapter 2 was written by H.Th. Jongen and the others by J. Guddat and F. Guerra.

We wish to express our thanks to all colleagues and students who directly or indirectly contributed to the making of this book, in particular to R. Lehmann (former student at the Humboldt University, Berlin), W. Zulehner (Kepler University, Linz), K. Wendler, D. Nowack, H. Günzel, R. Henrion, R. Schultz (Humboldt University, Berlin), M. Otero Pereira (Havana University) and J. Rückmann (TH Leipzig). We are particularly indebted to J. Stoer (University Würzburg) and L. Grippo (IASI, Rome), for valuable suggestions with respect to Section 6.2. Special thanks are due to L. Popova for her support with the list of references and to J. Kerger for her assistance in preparing the final version of the English manuscript. We would like to thank Ch. Dobers and S. Schmidt for their careful typing of the manuscript.

We further gratefully acknowledge the support of Mrs H. Ramsey, Miss I. Cooper and Dr P. Spuhler, extended in the name of the publishers.

We especially remark that comments and criticisms are always welcome; these should be directed to the first author at the following address: DDR-1086 Berlin, Humboldt-Universität, Sektion Mathematik, PSF 1297.

Aachen, Berlin, Havana, August 1989　　　　　　　　　　**J.G., F.G.V., H.Th.J.**

1

Introduction

1.1 A PRELIMINARY SURVEY ON SOLUTION ALGORITHMS IN ONE-PARAMETRIC OPTIMIZATION

We consider the following one-parametric optimization problem:

$$P(t): \qquad \min\{f(x,t)\,|\,x\in M(t)\}, \qquad t\in[0,1], \quad \text{resp.}\ t\in\mathbb{R}, \tag{1.1.1}$$

where

$$M(t) = \{x\subset\mathbb{R}^n\,|\,h_i(x,t)=0, i\in I, g_j(x,t)\leqslant 0, j\in J\} \tag{1.1.2}$$

with

$$I=\{1,\ldots,m\}, \quad m<n, \quad J=\{1,\ldots,s\}. \tag{1.1.3}$$

Throughout this book the functions f, h_i, g_j, $i\in I, j\in J$, are assumed to be k times continuously differentiable, where $k\geqslant 1$ will be specified later.

Now we introduce some well known notions that we want to use in this book (note that some of these notations are not unique in the literature). For details we refer to Chapter 2. The following system

$$D_x f(x,t) + \sum_{i\in I}\lambda_i D_x h_i(x,t) + \sum_{j\in J}\mu_j D_x g_j(x,t) = 0,$$

$$h_i(x,t)=0, \quad i\in I, \qquad g_j(x,t)\leqslant 0, \quad \mu_j\geqslant 0, \quad j\in J, \tag{1.1.4}$$

$$\mu_j\cdot g_j(x,t)=0, \quad j\in J$$

(cf. the glossary of symbols at the end of this book), is called a *Karush–Kuhn–Tucker system* (briefly *KKT system*). Each solution (x,λ,μ,t) of (1.1.4) is called at *KKT point* (sometimes the KKT point for $P(t)$ is denoted by (x,λ,μ) if it is clear that this point corresponds to $P(t)$, $z=(x,t)$ and x, respectively, are called a *stationary point*.

Of course, under some constraint qualifications the global and local minimizers z and x, respectively, are stationary points.

"

If we introduce the so-called *index set of active constraints* (briefly *active index set*) defined by

$$J_0(z) := \{ j \in J \mid g_j(z) = 0 \},$$

the KKT system (1.1.4) can be written in the following equivalent form:

$$D_x f(x,t) + \sum_{i \in I} \lambda_i D_x h_i(x,t) + \sum_{j \in J_0(z)} \mu_j D_x g_j(x,t) = 0, \tag{1.1.5a}$$

$$h_i(x,t) = 0, \quad i \in I, \qquad g_j(x,t) \leqslant 0, \quad j \in J, \tag{1.1.5b}$$

$$\mu_j \geqslant 0, \; j \in J_0(z). \tag{1.1.5c}$$

The following generalization of a stationary point plays an important role in our book. A point $z = (x,t) \in \mathbb{R}^n \times \mathbb{R}$ is called a *generalized critical point* (briefly g.c. point) if $x \in M(t)$ and the set

$$\{ D_x f(x,t), D_x h_i(x,t), D_x g_j(x,t), i \in I, j \in J_0(z) \}$$

is linearly dependent, i.e. there exist numbers $\lambda, \lambda_i, i \in I, \mu_j, j \in J_0(z)$, such that

$$\lambda D_x f(x,t) + \sum_{i \in I} \lambda_i D_x h_i(x,t) + \sum_{j \in J_0(z)} \mu_j D_x g_j(x,t) = 0,$$

$$h_i(x,t) = 0, \quad i \in I, \qquad g_j(x,t) \leqslant 0, \quad j \in J, \tag{1.1.6}$$

$$|\lambda| + \sum_{i \in I} |\lambda_i| + \sum_{j \in J_0(z)} |\mu_j| > 0.$$

Of course, for $\lambda \neq 0$ we can put $\lambda = 1$ and we get the KKT system under the additional condition (1.15c).

Now we can formulate the general aim of our book. Find a *solution algorithm* $P(t)$, $t \in [0,1]$. By a solution algorithm for $P(t)$, $t \in [0,1]$, we understand the solution of one of the following problems:

(A)   find a local minimizer $x(t)$ for $P(t)$, $t \in [0,1]$,

or

(B)   find a stationary point $x(t)$ for $P(t)$, $t \in [0,1]$,

or

(C)   find a g.c. point $x(t)$ for $P(t)$, $t \in [0,1]$;

more precisely, find a discretization (sufficiently fine)

$$0 = t_0 < \cdots < t_i < t_{i+1} < \cdots < t_N \tag{1.1.7}$$

of the interval $[0,1]$ and for each t_i, $i = 1, \ldots, N$,

(A)   a local minimizer $x(t_i)$,

or

(B) $\qquad\qquad\qquad$ a stationary point $x(t_i)$,

or

(C) $\qquad\qquad\qquad$ a g.c. point $x(t_i)$.

Of course, we assume that a local minimizer (stationary point or g.c. point) $x(0)$ as a starting point is known or easy to compute. In the case $P(t)$ has a global minimizer for all $t\in[0,1]$ (such an assumption will be necessary for several applications (cf. Section 1.2)), then $t_N = 1$ in (1.1.7).

There are many motivations to develop solution algorithms for $P(t)$, $t\in[0,1]$, e.g.

(i) solving parametric optimization problems arising in practical problems,

(ii) globally convergent algorithms (globalization of locally convergent algorithms by the approach of embedding),

(iii) global optimization,

(iv) multi-objective optimization,

(v) stochastic optimization (cf. e.g. Dupacova [42], Guddat *et al.* [83] and Kall [127]),

(vi) multi-level optimization problems, which especially appear in connection with decomposition methods (see e.g. Beer [15], Bank *et al.* [13] Geoffrion [66] and Tammer [218]),

(vii) semi-infinite optimization in connection with (finite-dimensional) multi-parametric optimization (cf. e.g. Hettich and Jongen [101, 102], Hettich and Still [103], Hettich and Zencke [104], Jongen *et al.* [120] and Jongen and Zwier [124–126]),

(viii) input optimization (cf. e.g. Zlobec [236–239]).

Section 1.2 includes an example of (i) and a brief survey of (ii), (iii) and (iv). Special results to (ii), (iii) and (iv) are included in Chapter 6. In particular, these applications show the difficulties in the non-convex case.

In order to develop solution algorithms we try to use information on $(x(t_i), t_i)$ to compute $(x(t_{i+1}), t_{i+1})$.

In the following we give a first answer to the following question: 'What kind of information could be helpful?' Of course, structure analysis and singularity theory of the following sets will be the theoretical basis for solution algorithms:

$$\Sigma_{gc} := \{(x, t)\in\mathbb{R}^n \times \mathbb{R} \mid x \text{ is a g.c. point for } P(t)\},$$

$$\Sigma_{stat} := \{(x, t)\in\mathbb{R}^n \times \mathbb{R} \mid x \text{ is a stationary point for } P(t)\},$$

4   *Parametric Optimization: Singularities, Pathfollowing and Jumps*

$$\Sigma_{KKT} := \{(x, \lambda, \mu, t) \in \mathbb{R}^n \times \mathbb{R}^m \times \mathbb{R}^s \times \mathbb{R} \,|\, (x, \lambda, \mu) \text{ is a KKT point for } P(t)\},$$

$$\Sigma_{loc} := \{(x, t) \in \mathbb{R}^n \times \mathbb{R} \,|\, x \text{ is a local minimizer for } P(t)\}.$$

We will see that for non-convex problems (which are of major interest in our book) Figure 1.1 is typical in the set Σ_{gc}. The exposed points in Figure 1.1 are points of different kinds of singularities. The notion of a singularity is closely related to the terms catastrophe and bifurcation (cf. e.g. Arnold *et al.* [8] Bröcker and Lander [22], Poore and Tiahrt [176] and Jongen *et al.* [121]).

The general theory for the sets Σ_{gc}, Σ_{stat}, Σ_{KKT} and Σ_{loc} was investigated by Dontchev and Jongen [41], Jongen *et al.* [112–118, 121], Kojima [134] and Kojima and Hirabayashi [135]. We refer in this context also to Pateva [173] for special classes, in particular linear optimization problems, and to Jongen *et al.* [119] and Schecter [206] for multiparametric optimization.

Continuity properties of curves in the sets introduced above are related to

(a) implicit function theorem results (cf. e.g. Fiacco [48, 49], Jongen *et al.* [122], Malanowski [149–152], Robinson [188–190, 192, 193, 195, 196], and the references cited there),

(b) qualitative and quantitative continuity of the optimal set mapping (cf. e.g. Bank *et al.* [12], Brosowski [24, 26], Fiacco and Kyparisis [50], Gfrerer [68], Klatte and Kummer [129], Klatte [130], Kummer [138], Malanowski [152], Zlobec and Ben–Israel [235], and the articles cited there),

(c) continuous selections (cf. e.g. Aubin and Cellina [9], Deutsch and Kenderov [35] Dommisch [39], Fischer [53], and the articles cited there).

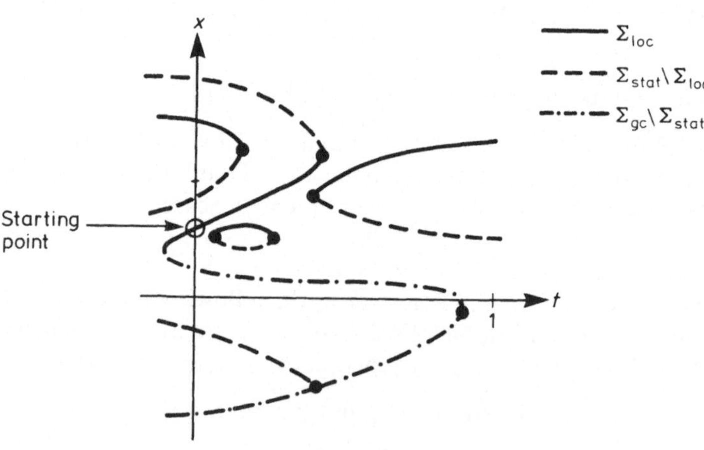

Figure 1.1

Furthermore, the following works from our bibliography belong to the theoretical background of solution procedures for parametric optimization problems: Armacost and Fiacco [7], Dontchev [40], Guddat and Jongen [86], Guddat *et al.* [87], Jongen and Zwier [124–126], Levitin [143], Robinson [197, 198] and Rupp [200, 201].

We may point out that the works of Fiacco [48, 49], Armacost and Fiacco [7] and Robinson [189] on the one hand and the article of Levitin [143] on the other provide the background for different notions of approximate selection functions (cf. Bank *et al.* [12], Section 6.3) and their computations.

A survey on further related investigations is included in Jongen and Weber [245].

The main tool for a solution algorithm will be the so-called pathfollowing methods (also called continuation methods or homotopy methods). For parameter-dependent nonlinear equations, pathfollowing methods are well known and successful (cf. e.g. Alexander *et al.* [1], Allgower and Georg [2, 3], Avila [11], Garcia and Zangwill [62], Georg [67], Ortega and Rheinboldt [169], Reinoza [177], Robinson [191], Schwetlick [209], and the works cited in all these articles and books, respectively).

There are two approaches to such methods (cf. Allgower and Georg [3] and the references cited there):

(1) Predictor–corrector methods (where an implicitly given continuous path P of solutions will be traced numerically),

(2) simplicial methods (where a piecewise linear approximation of the path P will be followed).

In our book we restrict ourselves to predictor–corrector methods and call them briefly pathfollowing methods. Now, there are several proposals for such methods developed directly for the optimization problem (e.g. Gfrerer *et al.* [70, 71] Guddat *et al.* [93] Hackl [97], Lehmann [140, 141], Meravy [161, 162], Richter [179, 180], Rupp [200, 2021] and Ruske [202, 203]). On the other

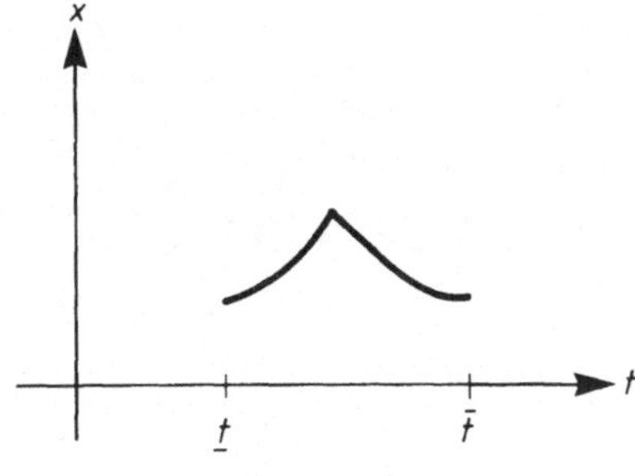

Figure 1.2

hand, we can reformulate the KKT system as a system of equations or a generalized equation and can use the predictor–corrector continuation methods cited above for parameter-dependent equations or generalized equations.

Now we want to describe the general idea of a predictor–corrector continuation method. For simplicity we assume the following condition:

(E1) There exists a continuous function $x:[\underline{t},\bar{t}]\to\mathbb{R}^n$, $\underline{t}<\bar{t}$, such that $x(t)$ is a local minimizer for all $t\in[\underline{t},\bar{t}]$ (cf. Figure 1.2).

From the structure analysis and the singularity theory mentioned above it follows that (E1) can be assumed, but we learn from Figure 1.1 that we cannot expect that $\underline{t}=0$ and $\bar{t}=1$ in the non-convex case. We want to find a discretization

$$\underline{t}=t_0<\cdots<t_i<t_{i+1}<\cdots<t_N=\bar{t} \tag{1.1.8}$$

and corresponding local minimizers $x(t_i)$, $i=1,\ldots,N$. More precisely, for an implementation on the computer, we search for a discretization (1.1.8) of $[\underline{t},\bar{t}]$, and for points $\tilde{x}^i$ $(i=1,\ldots,N)$ with $\|\tilde{x}^i-x(t_i)\|<\varepsilon$, $i=1,\ldots,N$, for a given $\varepsilon>0$ sufficiently small.

To solve this problem we consider such computation methods that generate $\tilde{x}^{i+1}$ in a small number of iteration steps with an efficient numerical effort using the previous point $\tilde{x}^i$ as a starting point. Here, we apply locally convergent algorithms with an at least superlinear or quadratic rate of convergence (cf. e.g. the methods of Garcia-Palomares and Mangasarian [61] Robinson [187] and Wilson [229]; these methods are described in Chapter 3). Methods of this type have proved their value on the computer, too (cf. Hock and Schittkowski [106] and Schittkowski [208]).

For convergence, these methods require the important assumption that the chosen starting point has to belong to a certain neighbourhood of the unknown local minimizer searched for. Thus, the idea described above can be realized by using the known point $(\tilde{x}^i,t_i)$ as an approximation of $(x(t_i),t_i)$ to compute

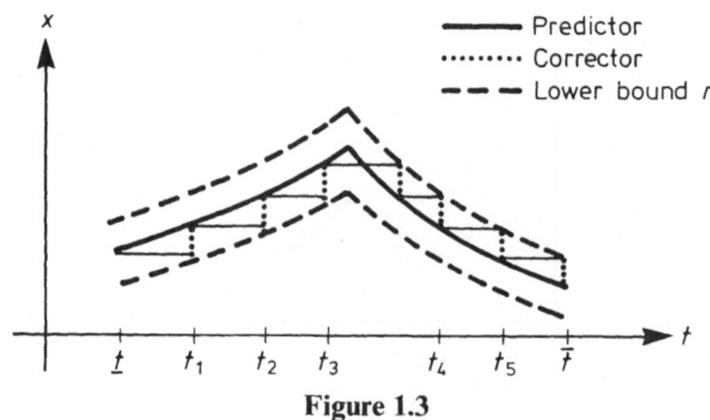

Figure 1.3

an approximation $(\tilde{x}^{i+1}, t_{i+1})$ of $(x(t_{i+1}), t_i)$ in an efficient way (cf. Figure 1.3) if $|t_{i+1} - t_i|$ is chosen sufficiently small.

In the literature we can distinguish three general approaches to realize this idea.

First approach

(Using locally convergent algorithms for the full problem $P(t)$.) We follow here Lehmann [140] and assume (E1). Now we consider a locally convergent algorithm (e.g. the methods mentioned above). Let $r(t)$ be the radius of convergence for solving $P(t)$ in a neighbourhood of $x(t)$. Then, under some additional assumptions, there is a positive number r such that

$$r(t) \geqslant r \qquad \text{for all} \qquad t \in [\underline{t}, \bar{t}] \tag{1.1.9}$$

(cf. Figure 1.3). For details the reader is referred to Chapter 3.

Of course, the estimation (1.1.9) is another kind of information to realize the transition $(x, (t_i), t_i) \rightarrow (x(t_{i+1}), t_{i+1})$.

Second approach

(Transform the KKT system to a system of equations or to a generalized equation, and use pathfollowing methods for nonlinear equations or generalized equations.) Both transformations are well known. We point out a system of the transformation of the KKT system to a system of equations. This is standard in nonlinear complementarity theory and also in nonlinear optimization (cf. e.g. Eaves and Scarf [45] for the linear case; Charnes *et al.* [31], Garcia and Zangwill [62], Kojima [132], Kojima and Hirabayashi [135], Megiddo and Kojima [160] and Mangasarian [153]. Of course, the KKT system (1.1.4) is a mixture of equations and inequalities. The reformulation to a system of equations is quite simple. For $y \in \mathbb{R}$ we define

$$y^+ := \max\{y, 0\}, \qquad y^- := \min\{y, 0\}.$$

We can formulate the KKT system as an equivalent system of equations:

$$\mathcal{H}(x, y, t) := \begin{bmatrix} D_x f(x, t) + \sum_{i \in I} y_i D_x h_i(x, t) + \sum_{j \in J} y_j^+ D_x g_j(x, t) \\ h_i(x, t), \quad i \in I \\ y_j^- - g_j(x, t), \quad j \in J \end{bmatrix} = 0. \tag{1.1.10}$$

More precisely, if (x, λ, μ) satisfies the KKT system (1.1.4), then (x, y) solves (1.1.10), where $y_i = \lambda_i$, $i \in I$, and for $j \in J$

$$y_j = \begin{cases} \mu_j & \text{if } g_j(x, t) = 0, \\ g_j(x, t) & \text{if } g_j(x, t) < 0. \end{cases}$$

Conversely, if (x, y) satisfies (1.1.10), then (x, λ, μ) fulfils the KKT system, where

$$\lambda_i = y_i, \quad i \in I, \qquad \mu_j = y_j^+, \quad j \in J, \qquad g_j(x, t) = y_j^-, \quad j \in J.$$

Then (cf. Kojima and Hirabayashi [135], Section 2.6), the mapping $\mathcal{H}$ is piecewise continuously differentiable (shortly, PC^1) with respect to the subdivision $\{\tau(K) | K \subseteq J\}$ of $\mathbb{R}^{n+m+s+1}$, where the so-called cells $\tau(K)$ are defined by

$$\tau(K) = \mathbb{R}^n \times \{y \in \mathbb{R}^{m+s} | y_i \geqslant 0, i \in K, y_j \leqslant 0, j \in J \backslash K\} \times \mathbb{R},$$

i.e. for each $\tau(K)$ there exists an open set $U \supseteq \tau(K)$ and a continuously differentiable function $\mathcal{G}: U \to \mathbb{R}^{n+m+s}$ with $\mathcal{G}|_{\tau(K)} = \mathcal{H}|_{\tau(K)}$. Then we can use pathfollowing methods for this PC^1 mapping (cf. e.g. Alexander *et al.* [1] and Reinoza [177]).

There are also possibilities to reformulate the KKT system into a k-times ($k \geqslant 1$) continuously differentiable nonlinear parameter-dependent system of equations (cf. e.g. Garcia and Zangwill [62] and Mangasarian [153]). Then we can use standard methods for parameter-dependent equations (cf. the literature mentioned above). Furthermore, we can also consider the KKT system as a generalized equation (cf. e.g. Robinson [190, 192, 193]) and use the corresponding pathfollowing methods (cf. e.g. Reinoza [177] and Richter [180]).

We do not follow the second approach in our book and refer the interested reader to the cited literature.

The first two approaches have the disadvantage that we have to use the full KKT system or the corresponding system of generalized equations. This disadvantage is overcome in the next approach.

Third approach

(Active index set strategy.) Instead of the original problem $P(t)$, $t \in [\underline{t}, \bar{t}]$, we consider a finite number of auxiliary parametric optimization problems with equality constraints of the following type only:

$$P^{J_s}(t): \qquad \min \{f(x, t) | h_i(x, t) = 0, i \in I, g_j(x, t) = 0, j \in J_s\}, \qquad t \in [t_s, t_{s+1}],$$

where

$$J_s \subseteq J, t_s < t_{s+1}, s = 0, 1, \ldots, N-1, t_0 = \underline{t}, t_N = \bar{t}.$$

This has the advantage of reducing the size of the problem since the sets J_s are active index sets. Moreover, since $P^{J_s}(t)$ has only equality constraints, we can use the predictor–corrector schemes for solving the nonlinear equations mentioned above (the corresponding KKT system is a nonlinear equation system). However, the points where the active index sets will be changed are singularities and have to be overcome, and the new index set of active constraints has to be computed.

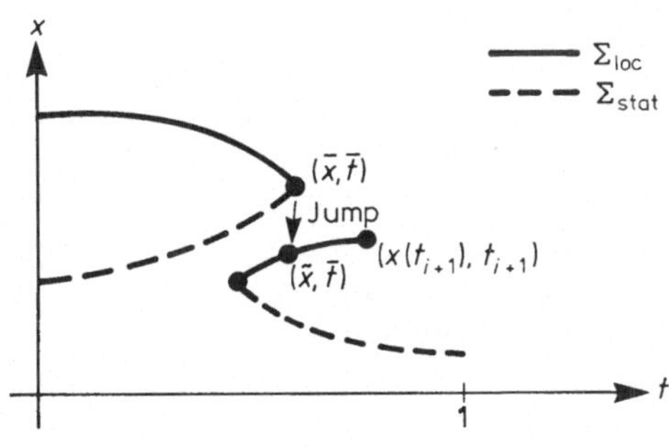

Figure 1.4

Chapter 3 and Sections 4.2 and 4.3 contain pathfollowing methods in the set Σ_{stat} and Σ_{KKT}, respectively, based on [70], [71] and [93]. As we see in Figure 1.4 the path in Σ_{stat} stops at a point $z = (x, t)$ although there is a continuation in Σ_{gc} ($\bar{z}$ is another kind of singularity). For this reason we extend the pathfollowing methods to connected components in Σ_{gc} (cf. Section 4.4). There is no doubt that the pathfollowing methods are a powerful technique to follow a connected component in Σ_{loc}, Σ_{stat} and Σ_{gc} numerically, but we cannot expect to reach $t_N = 1$ by using pathfollowing methods only (cf. Figure 1.4). For this reason we propose to jump to another connected component in Σ_{loc} (Section 5.2) and Σ_{gc} (Section 5.3), respectively, as shown in Figure 1.4 for Σ_{loc}.

In some cases a jump can be realized by using the information on the singularities. The transition from $(x(t_i), t_i)) \rightarrow (x(t_{i+1}), t_{i+1})$ will be realized in the following way:

$$(x(t_i), t_i) = (\bar{x}, \bar{t}) \rightarrow (\tilde{x}, \tilde{t}) \rightarrow (x(t_{i+1}), t_{i+1}).$$

Since we do not have proposals for jumps in all cases, we cannot guarantee to reach $t = 1$. This fact is not unexpected if we have a look at the applications (globally convergent algorithms, global optimization and multi-objective optimization) worked out in Chapter 6.

Finally, we want to mention that there exist special pathfollowing methods for linear optimization problems with one parameter in the objective and in the right-hand sides, convex quadratic optimization problems with one parameter in the linear part of the objective and in the right-hand sides of the linear constraints and a special complementarity problem for which it is possible to compute a vector function $x(t)$ defined on $[0, 1]$ under the assumption that $P(t)$ has a global minimizer for all $t \in [0, 1]$. These methods are not the subject of our book, but we give a short outline. We can use modified pivot algorithms for the special classes above. Such methods are described in several books (e.g. Bank *et al.* [12], Dinkelbach [38], Gal [57, 58], Guddat *et al.* [83], Nožička *et al.* [168] and van de Panne [170]). We note that the computation of $x(t)$ on

[0, 1] is really efficient. These procedures, however, were founded on the inner rules of the pivot method. In particular, the basic idea for the methods for convex quadratic optimization problems with one parameter in the linear part of the objective and one parameter in the right-hand side is originally due to Wolfe [230], who, in the third phase of his method for solving convex quadratic optimization problems, found it necessary to solve a special one-parametric quadratic auxiliary problem. The same basic idea underlies the well known procedure of Lemke [142] for solving linear complementary problems in which a special one-parametric linear complementarity problem is involved in a subsidiary rule. Other authors have shown that the procedure of Lemke and, consequently, as its central point the procedure for solving the auxiliary parametric problem, always provide a solution in a larger class of matrices than that considered by Lemke cf. e.g. Chandrasekaran [29] and Saigal [205]). Detailed expositions are given by Lüthi [148] and van de Panne [170]. What can we say about other classes of one-paraametric optimization problems? The monograph of van de Panne [170] contains a description of a procedure for solving one-parametric complementarity problems in general form. We also refer to works of Grygerova [79] and Gollmer [73] and the articles cited there on linear optimization problems with parameter-dependent constraint matrices. A procedure based on the pivot technique and designed to solve certain one-parametric piecewise linear complementarity problems is that of Kojima *et al.* [136]. This also applies to a procedure of Kojima [132], which determines a piecewise linear approximation for a continuously parameter-dependent KKT curve of a not necessarily convex one-parametric optimization problem. Proposals for dealing with one-parametric nonlinear complementarity problems were made Megiddo [159]. Further we want to mention a procedure of Zsigmond [241] for linear problems with one parameter in the contraint matrix as well as an algorithm of Geoffrion [65] for solving strictly convex one-parametric problems entailing analytic functions. In the case of quadratic problems in which the parameter occurs in the objective function matrix too, there also exist proposals for solution procedures (cf. e.g. Boot [21], Ritter [183], Guddat and Tammer [92], Tammer [215] and Välialo [221]). However, studies by Guerra [94] confirm that most of the procedures mentioned before are hardly suitable for computing a discretization (1.1.7) and corresponding solutions $x(t_i)$, $i = 1, \ldots, N$, for large problems. For such classes it will be more successful to use the nonlinear approach proposed by Pateva [173]).

1.2 SOME MOTIVATIONS

This section includes some different applications of solution algorithms for one-parametric optimization problems. The first example is a practical example.

The second (globally convergent algorithms for nonlinear optimization and global optimization) and the third (multi-objective optimization) examples show the main difficulties in finding a solution algorithm in the non-convex case. Globally convergent algorithms, global optimization and multi-objective optimization will be investigated deeper in Chapter 6. It becomes obvious that these fundamental problems could be solved if we found a solution algorithm for one-parametric optimization problems. Nevertheless, the investigation described in Chapter 6 provides a deeper insight into these important problems.

1.2.1 A parametric model for the optimal dispatch of thermal power stations within short time

The following model (cf. also Huneault *et al* [108]) can be used for the optimization of costs for the load generation in a system of thermal power stations within short time, e.g. less than one hour. The whole load generation system may additionally involve other modes of generation like hydroelectric power stations and pumped storage plants, but here we assume the operation of these modes to be fixed in such a way that the system including the thermal plants is able to meet the total load demand.

Assume that there are N operating thermal units with capacities bounded above and below. Let the variable x_i denote the unknown load for the ith unit and $\underline{x}_i, \bar{x}_i$ its capacity bounds ($i = 1, \ldots, N$). Then we have

$$\underline{x}_i \leqslant x_i \leqslant \bar{x}_i, \qquad i = 1, \ldots, N. \tag{1.2.1}$$

Let $a_i(x_i)$ be the (given) cost function for the ith unit. Then the total costs for load generation will be

$$\sum_{i=1}^{N} a_i(x_i). \tag{1.2.2}$$

Furthermore, denote

$$P_{\min} := \sum_{i=1}^{N} \underline{x}_i \qquad \text{and} \qquad P_{\max} := \sum_{i=1}^{N} \bar{x}_i.$$

Then, the assumption that the operating units are able to meet the current load demand means that there exists a $t \in [0, 1]$ such that the demand is given by

$$P_{\min} + t(P_{\max} - P_{\min}).$$

We obtain the following balance equation between generation and demand

$$\sum_{i=1}^{N} x_i = P_{\min} + t(P_{\max} - P_{\min}), \qquad t \in [0, 1]. \tag{1.2.3}$$

Here $t \in [0, 1]$ is considered a parameter corresponding to different values of the current demand.

Now, (1.2.1), (1.2.2) and (1.2.3) lead to the following one-parametric optimization problem

$$P(t) \quad \min\left\{ \sum_{i=1}^{N} a_i(x_i) \,\middle|\, \underline{x}_i \leqslant x_i \leqslant \bar{x}_i, i = 1, \ldots, N, \sum_{i=1}^{N} x_i = P_{\min} + t(P_{\max} - P_{\min}) \right\},$$

$$t \in [0, 1].$$

We note that in the case when the objective function $\sum_{i=1}^{N} a_i(x_i)$ is strictly convex and quadratic, (E1) is fulfilled with $\underline{t} = 0$ and $\bar{t} = 1$. Moreover, there exists a unique solution $x(t)$ for all $t \in [0, 1]$ and x is a piecewise linear function (cf. e.g. Bank *et al.* [12]). In this case we can us a simple procedure using the special structure (cf. Kleinmann and Schultz [131]). In the case when the objective function does not have such nice properties, we propose to use more general pathfollowing methods.

Furthermore, we observe that the set of active constraints changes in a finite number of $t_i \in (0,1)$ and more than one index might change at such a parameter value t_i (cf. Chapter 3).

1.2.2 On globally convergent algorithms and global optimization

We consider the following nonlinear optimization problem

$$\text{(P)} \qquad \min\{ f(x) \,|\, h_i(x) = 0, i \in I, g_j(x) \leqslant 0, j \in J \} \qquad (1.2.4)$$

where $I = \{1, \ldots, m\}$, $m < n$, $J = \{1, \ldots, s\}$, and $f, h_i, g_i \in C^2(\mathbb{R}^n, \mathbb{R})$, $i \in I$, $j \in J$.

We turn to the problem of finding at least one g.c. point of (P) without any additional information like

(i) convexity (in this case, for example, penalty and Lagrangian methods are convergent with an arbitrary starting point, cf. e.g. [78]), or

(ii) a point in a sufficiently small neighbourhood of a stationary point is known (locally convergent algorithms mentioned above), or,

(iii) a point in a sufficiently small neighbourhood of an 'optimal' Lagrange vector is known (so-called augmented Lagrangian methods, cf. e.g. [18]).

Thus, we consider the problem of finding a local minimizer and a stationary point or a g.c. point of the non-convex optimization problem (P) using an arbitrary starting point $x^0 \in \mathbb{R}^n$. We call such a method a globally convergent algorithm.

If we were able to find such an algorithm, we would also have a deterministic approach for solving the problem of global optimization. Up to now there does not exist a really powerful deterministic approach to solve this problem. Now we introduce the problem of global optimization. Let $F \in C^2(\mathbb{R}^n, \mathbb{R})$ be a given function. We consider the problem:

find a global minimizer of $F \in C^2(\mathbb{R}^n, \mathbb{R})$ subject to K,
where K is a simple convex and compact subset of $\mathbb{R}^n$,
e.g. $K := \{x \in \mathbb{R}^n | \, \|x\|^2 \leqslant q\}$ and q is a given positive number.

This problem is symbolized by

$$\text{glob min} \{F(x) | x \in K\}. \tag{1.2.5}$$

We propose the following simple strategy (cf. Figure 1.5):

Step 1: Compute a stationary point $\hat{x}$ for

$$\min \{F(x) | x \in K\}.$$

Step 2: Find a point belonging to

$$\{x \in K | F(x) \leqslant F(\hat{x}) - \varepsilon\},$$

where $\varepsilon > 0$ is sufficiently small, by finding a g.c. point of

$$\min \{\|x - x^0\|^2 | x \in K, F(x) \leqslant F(\hat{x}) - \varepsilon\}, \tag{1.2.6}$$

with $x^0 \in K$ arbitrarily chosen.

Of course, step 2 is the difficult one. This step can be realized if we have a globally convergent algorithm for the problem (1.2.6). We note that $\hat{x}$ is a global minimizer for the problem $\min \{F(x) | x \in K\}$ if and only if $\{x \in K | F(x) \leqslant F(\hat{x}) - \varepsilon\} = \varnothing$.

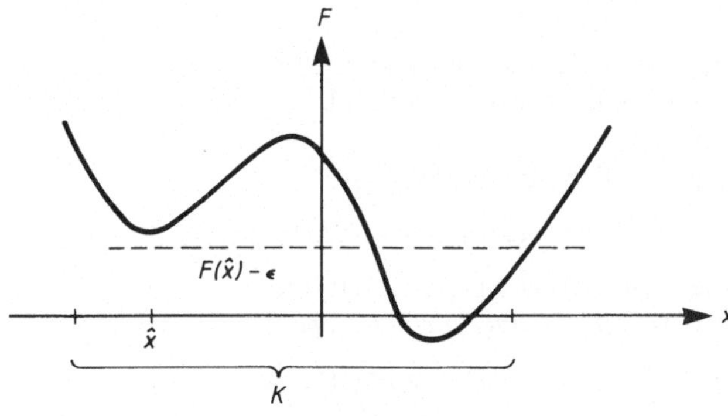

Figure 1.5

Now we return to the general problem (P) defined in (1.2.4). We propose the well known concept of embedding. Choose a one-parametric optimization problem $P(t)$, $t \in [0, 1]$, of the form (1.1.1) with at least the following properties:

(E2) A local minimizer x^0 of $P(0)$ is known,

(E3) $P(1) = (P)$.

The conditions (E2) and (E3) are fulfilled if we choose e.g. $P(t)$ defined by

$$f(x, t) := tf(x) + (1 - t)\|x - x^0\|^2, \tag{1.2.7}$$

$$h_i(x, t) := h_i(x) + (t - 1)h_i(x^0), \qquad i \in I, \tag{1.2.8}$$

$$g_j(x, t) := g_j(x) + (t - 1)|g_j(x^0)|, \qquad J \in J, \tag{1.2.9}$$

where $x^0 \in \mathbb{R}^n$ is arbitrarily fixed. We note that x^0 is even a global minimizer.

Of course, a solution algorithm for $P(t)$ with (1.2.7), (1.2.8) and (1.2.9) will be globally convergent for (P). We note that the pathfollowing methods described in Chapters 3 and 4 will be successful if e.g. condition (E1) is satisfied for $t = 0$ and $\bar{t} = 1$. But, as remarked in Section 1.1 (cf. Figure 1.1 and the investigation in Section 6.2), this assumption is not fulfilled in general. We refer the reader to Sections 6.2 and 6.3 for deeper information.

1.2.3   Multi-objective optimization based on parametric optimization

We consider the following multi-objective optimization problem

$$\min \{(f_1(x), \ldots, f_L(x)) | x \in M\},$$

where M is a given non-empty subset of $\mathbb{R}^n$ and $f_1, \ldots, f_L$ are given functions defined on $\mathbb{R}^n$.

We use here the well-known notions of an efficient point (cf. e.g. Pareto [172] and Kuhn and Tucker [137] (also called Pareto-optimal, admissible, non-dominated, etc.) and a locally efficient point, and the less well known notions of a properly efficient point with the bound ε (cf. Wierzbicki [227]) and a weakly efficient point (cf. Brosowski and Conci [25]). We remind the reader of these notion.

A point $\bar{x} \in M$ is called an *efficient point* if

$$(f(\bar{x}) + D) \cap f(M) = \emptyset,$$

where $D = -\mathbb{R}^L_+ \backslash \{0\}$, $f(x) = (f_1(x), \ldots, f_L(x))$.

A point $\bar{x} \in M$ is called *properly efficient point* with the bound ε if

$$(f(\bar{x}) + \tilde{D}_\varepsilon) \cap f(M) = \emptyset,$$

where $\tilde{D}_\varepsilon = D_\varepsilon \backslash \{0\}$, $D_\varepsilon = \{y \in \mathbb{R}^L | \text{dist}(y, -\mathbb{R}^L_+) \leqslant \varepsilon\}$.

A point $\bar{x} \in M$ is called a *weakly efficient point* if

$$(f(\bar{x}) + \tilde{D}) \cap f(M) = \varnothing,$$

where $\tilde{D} = -\operatorname{int} \mathbb{R}^L_+$.

By M_{eff} we denote the set of all efficient points; by M_{loceff} the set of all locally efficient points; by $M^\varepsilon_{\text{eff}}$ the set of all properly efficient points with bound ε; by $M^\varepsilon_{\text{loceff}}$ the set of all locally properly efficient points with bound ε; by M_{weff} the set of all weakly efficient points; and by M_{locweff} the set of all locally weak efficient points. Then we have:

$$M^\varepsilon_{\text{eff}} \subset M_{\text{eff}} \subset M_{\text{eff}}.$$

We consider the following three parametrizations ($\mu \in \mathbb{R}^L$ is always the parameter vector).

First parametrization

We consider the following objective function (cf. Wierzbicki [227]:

$$s(f(x), \mu) = \max_{i \in \{1,\dots,L\}} \lambda^0_i(f_i(x) - \mu_i) + \delta \sum_{i=1}^L \lambda^0_i(f_i(x) - \mu_i), \tag{1.2.10}$$

where $\lambda^0_i > 0$, $i = 1, \dots, L$, are fixed and $\delta \in (0, 1)$ is fixed with $\delta < \varepsilon$.

Let $\psi_1(\mu)$ ($\psi_{1,\text{loc}}(\mu)$) be the set of all global (local) minimizers for

$$\min\{s(f(x), \mu) \mid x \in M\}, \qquad \mu \in \mathbb{R}^L. \tag{1.2.11}$$

Then (1.2.11) has the property

$$\bigcup_{\mu \in \mathbb{R}^L} \psi_1(\mu) \subset M_{\text{eff}}, \qquad \left(\bigcup_{\mu \in \mathbb{R}^L} \psi_{1,\text{loc}}(\mu) \subset M_{\text{loceff}}\right). \tag{1.2.12}$$

The closure of the left-hand side in (1.2.12) tends to M_{eff} as $\varepsilon \to 0$:

$$M_{\text{eff}} \subset \bigcup_{\mu \in f(M_{\text{eff}})} \{x \in \psi_1(\mu) \mid s(f(x), \mu) = 0\} \subset \bigcup_{\mu \in \mathbb{R}^L} \psi_1(\mu), \tag{1.2.13}$$

$$\left(M_{\text{loceff}} \subset \bigcup_{\mu \in f(M_{\text{loceff}})} \{x \in \psi_{1,\text{loc}}(\mu) \mid s(f(x), \mu) = 0\} \subset \bigcup_{\mu \in \mathbb{R}^L} \psi_{1,\text{loc}}(\mu)\right)$$

(cf. Theorem 10 in [227]).

We note that $s(f(\cdot), \mu)$ defined by (1.2.10) is not differentiable and, therefore, not suitable for pathfollowing methods. However, by taking an additional variable v, it is possible to transform (1.2.11) into a differentiable problem

$$P_1(\mu): \quad \min\left\{\delta \sum_{i=1}^L \lambda^0_i(f_i(x) - \mu_i) + v \,\middle|\, x \in M, \lambda^0_i f_i(x) - v \leqslant \lambda^0_i \mu_i, i = 1, \dots, L\right\},$$

$$\mu \in \mathbb{R}^L. \tag{1.2.14}$$

Second parametrization

$$P_2(\mu): \qquad \min\left\{\frac{1}{\rho}\ln\left[\frac{1}{L}\sum_{i=1}^{L}\lambda_i^0\left(\frac{\tilde{q}_i-f_i(x)}{\tilde{q}_i-\mu_i}\right)^\rho\right]\Bigg|\, x\in M\right\}, \qquad (1.2.15)$$

where

$$\tilde{q}_i < \min_{x\in M} f_i(x), \qquad i=1,\ldots,L, \rho \geqslant L$$

(cf. Grauer *et al.* [76]).

By $\psi_2(\mu)$ ($\psi_{2,\mathrm{loc}}(\mu)$) we denote the set of all global (local) minimizers. Then we have

$$\bigcup_{\mu\in\mathcal{M}}\psi_2(\mu)\subset M_{\mathrm{eff}}, \qquad \left(\bigcup_{\mu\in\mathcal{M}}\psi_{2,\mathrm{loc}}(\mu)\subset M_{\mathrm{loceff}}\right) \qquad (1.2.16)$$

with $\mathcal{M}=\{\mu\in\mathbb{R}^L\,|\,\mu_i>\tilde{q}_i,\ i=1,\ldots,L\}$, and

$$M_{\mathrm{eff}}^\varepsilon \subset \bigcup_{\mu\in f(M_{\mathrm{eff}}^\varepsilon)\cap\mathcal{M}}\{x\in\psi_3(\mu)\,|\,\tilde{s}(f(x),\mu)=0\}\subset\bigcup_{\mu\in\mathcal{M}}\psi_2(\mu), \qquad (1.2.17)$$

$$\left(M_{\mathrm{loceff}}^\varepsilon \subset \bigcup_{\mu\in f(M_{\mathrm{loceff}})\cap\mathcal{M}}\{x\in\psi_{2,\mathrm{loc}}(\mu)\,|\,\tilde{s}(f(x),\mu)=0\}\subset\bigcup_{\mu\in\mathcal{M}}\psi_{2,\mathrm{loc}}(\omega)\right),$$

where $\tilde{s}(f(\cdot),\mu)$ denotes the objective function in (1.2.14).

Third parametrization

$$P_3(\mu):\min\{v\,|\,(x;v)\in M_3(\mu)\}, \qquad \mu\in\mathbb{R}^L, \qquad (1.2.18)$$

where

$$M_3(\mu):=\{(x,v)\in\mathbb{R}^n\times\mathbb{R}\,|\,x\in M, f_j(x)-v\leqslant\mu_j, j=1,\ldots,L\}$$

(cf. Wierzbicki [227]).

By $\tilde{\psi}_3(\mu)$ ($\tilde{\psi}_{3,\mathrm{loc}}(\mu)$) we denote the set of all global (local) minimizers, and define

$$\psi_3(\mu):=\{x\in\mathbb{R}^n\,|\,\exists v\in\mathbb{R}:(x,v)\in\tilde{\psi}_3(\mu)\},$$
$$(\psi_{3,\mathrm{loc}}(\mu):=\{x\in\mathbb{R}^n\,|\,\exists v\in\mathbb{R}:(x,v)\in\tilde{\psi}_{3,\mathrm{loc}}(\mu)\}).$$

Then we have

$$\bigcup_{\mu\in\mathbb{R}^L}\psi_3(\mu)=M_{\mathrm{weff}} \qquad \left(\bigcup_{\mu\in\mathbb{R}^L}\psi_{3,\mathrm{loc}}(\mu)=M_{\mathrm{locweff}}\right) \qquad (1.2.19)$$

(cf. Broswski and LConci [25]).

The relations (1.2.12), (1.2.13), (1.2.16), (1.2.17) and (1.2.19) show that we can use the parametric optimization problems $P_i(\mu)$, $i=1,2,3$, for the computation of points interesting for the decision-maker. Of course, there are other parametrizations (cf. e.g. Charnes and Cooper [30], Gal [59], Guddat *et al.* [83], Ester and Tröltzsch [47] and Wierzbicki [227]).

We have to note the parametrizations $P_i(\mu)$, $i = 1, 2, 3$, are multiparametric optimization problems. In our approach (cf. also Guddat *et al.* [83]) we reduce the multiparametric optimization problem $P_i(\mu)$, $i \in \{1, 2, 3\}$, to a sequence of one-parametric optimization problems, which is generated by a dialogue procedure. This can be obtained by taking suitable points μ^0 and μ^1 and the line segment $\{\mu \in \mathbb{R}^L | = \mu^0 + t(\mu^1 - \mu^0), \ t \in [0, 1]\}$ connecting the two points μ^0 and μ^1. This implies one-parametric optimization problems

$$P_i(t) := P_i(\mu^0 + t(\mu^1 - \mu^0)), \qquad t \in [0, 1], \ i = 1, 2, 3.$$

In the following we explain more precisely how we obtain the points μ^0 and μ^1. Let x^k be a currently computed efficient (resp. locally efficient, etc.) point. Then we consider a telescreen picture described in Table 1.1. The main information consists of estimating the objective function values of the point x^k in comparison to $\underline{f}_i = \inf\{f_i(x) | x \in M\}$, $i = 1, \ldots, L$.

The third column contains the percentage of deviations of the current objective function values $f_i(x^k)$ from the lower bounds $\underline{f}_i$. Clearly, these quantities have to be given suitable values in the cases of $\underline{f}_i = 0$ and $\underline{f}_i = -\infty$. When the problem $\min\{f_i(x) | x \in M\}$ is solvable, a good approximation of $\underline{f}_i$ can only be made in the convex case, in general. Otherwise only local minima are obtained. However, the decision-maker should answer the following questions by using this telescreen picture:

(a) Which $f_i (i \in \{1, \ldots, L\})$ do you wish to improve? Let $K \subset \{1, \ldots, L\}$ be the corresponding index set.

(b) Which goals a_j do you wish for f_i, $i \in K$?

(c) Which upper bounds a_j can you accept for f_j, $j \in \{1, \ldots, L\} \backslash K$?

In answering these three questions the decision-maker can express his current wishes. The intended dialogue control in the sense of realizing these wishes can be achieved by choosing $\mu_i^1 = a_i (j = 1, \ldots, L)$. Here $a_j = +\infty$ means that the corresponding inequality $f_j(x) \leqslant \mu_j$ must be deleted. The starting parameter μ^0 will be chosen by $f_i(x^k) \leqslant \mu_j^0$, $i = 1, \ldots, L$. One main problem in multi-objective

Table 1.1

$\underline{f}_1$	$f_1(x^k)$	$\dfrac{f_1(x^k) - \underline{f}_1}{|\underline{f}_1|} \times 100$
$\underline{f}_i$	$f_i(x^k)$	$\dfrac{f_i(x^k) - \underline{f}_i}{|\underline{f}_i|} \times 100$
$\underline{f}_L$	$f_2(x^k)$	$\dfrac{f_L(x^k) - \underline{f}_L}{|\underline{f}_2|} \times 100$

optimization will be the answer to the question whether the goal μ^1 was a realistic one or not. Furthermore, the decision-maker is interested in finding a point realizing a realistic goal. We refer the reader to Section 6.4 for further information.

1.3 SUMMARIES OF CHAPTERS 2–6

Chapter 2

In this chapter we present a compilation of several results in parametric optimization that are useful for the development of computational methods on the one hand, and provide a general understanding on the other. Special attention is paid to one-parametric optimization problems, the critical curves involved and the types of singularities occurring generically.

First, we will consider unconstrained optimization problems. The structure of the set Σ_{gc}, in particular singularities, will be discussed. Secondly, the feasible set will be investigated. Here, two constraint qualifications (the so-called linear independence constraint qualification and the Mangasarian–Fromovitz constraint qualification) play an important role. The latter constraint qualification is a necessary property for the feasible set to be stable.

Thirdly, the class $\mathscr{F}^{**}$ of Jongen, Jonker and Twilt will be introduced. This is a generic class where Σ_{gc} is divided into five types (the point of type 1 is non-degenerate, the others represent different kinds of singularities). This class together with the class of Kojima and Hirabayashi (zero is a regular value of the corresponding KKT mapping $\mathscr{H}$ (cf. (1.1.10)) will be the theoretical basis for the pathfollowing methods described in Chapters 3 and 4. Furthermore, the class $\mathscr{F}^{**}$ can be used for constructing jumps in the sets Σ_{loc} and Σ_{gc} (cf. Chapter 5). We note that there exists also a classification of singularities for the class of Kojima and Hirabayashi (cf. Rückmann [199] and Guddat *et al.* [89]) and for linear optimization problems (cf. Pateva [173]). The latter two investigations are not included in our book.

Chapter 3

First, the radius of convergence depending on the parameter t for a general locally convergent algorithm (including the algorithms of Robinson, of Wilson and of Garcia-Palomares and Mangasarian, cf. [229], [187], [61]) will be estimated by a positive constant under some assumptions. This is a generalization of the results given by Avila for one-parameter dependent equations.

Secondly, an active index set strategy will be proposed to follow a function of local minimizers numerically. Here, in particular, the changing of more than one index is allowed. This investigation will lead to the algorithm PATH I, which works in the set Σ_{loc}.

Chapter 4

In distinction to Chapter 3, we will also consider connected components in Σ_{stat} and Σ_{gc}. The algorithm PATH II works in Σ_{stat} and needs the assumption that zero is a regular value of the KKT mapping $\mathscr{H}$ (cf. (1.1.10)). The main point of the algorithm PATH III (in the set Σ_{gc}) lies in the construction of the active index sets by using information on the singularities. This means that we will consider the class $\mathscr{F}^{**}$. This algorithm has the advantage that we can continue a path in $\Sigma_{\mathrm{gc}} \backslash \Sigma_{\mathrm{stat}}$ (cf. Figure 1.1), which is a useful extension of the classical pathfollowing methods.

Chapter 5

First, jumps from one connected component in Σ_{loc} to another one will be proposed if different types of turning points appear. Here, we can use the information on the singularities in the class $\mathscr{F}^{**}$ in some cases to find a direction of descent. The possible jumps will be included in the algorithm JUMP I. Unfortunately, we do not have a proposal for a jump in all cases. That is the reason why we will consider, secondly, also jumps in the set Σ_{gc}. Here, we aim to describe numerically as many connected components in Σ_{gc} as possible.

This will lead to the algorithm JUMP II. The algorithm cannot guarantee that we will find all connected components, but it extends the possibilities for jumps in comparison to the algorithm JUMP I.

Chapter 6

Here we will include applications to:

(a) Globally convergent algorithm for nonlinear optimization problems by the concept of embedding.

(b) global optimization,

(c) multi-objective optimization.

We refer to Section 1.2 for summary. The main result of this chapter consists of the better understanding of the difficulties of these three problems. We can propose partial solutions only. This is due to the fact that we have to find a numerical description of all connected components in Σ_{gc} in the worst case, or in other words, we need jumps for all situations that could appear. This is still an open problem. This result is not unexpected because the problems (a), (b) and (c) are solved, restricted to the set $\mathscr{F}^{**}$, if we can develop a solution algorithm for $P(t)$, $t \in [0, 1]$, which works in any case. From this point of view the problems in (a), (b) and (c) have the same degree of difficulty.

2

Theoretical Background

(by H. Th. Jongen)

2.1 PRELIMINARY OUTLINE

In this chapter we present a compilation of several results in parametric optimization that are useful for the development of computational methods on the one hand and provide a general understanding on the other. In view of the scope of this book, we focus our attention on finite-dimensional differentiable optimization problems depending on one real parameter. For details on the proof of the results we refer to the cited literature; this gives us the possibility to pay more attention to additional clarifying remarks. Specific difficulties arising in multiparametric problems are discussed in [206] and [119].

The organization of the theoretical background is as follows. In Section 2.2 we start with unconstrained optimization problems. Section 2.3 is devoted to constraint sets, their local structure and constraint qualifications. In Section 2.4 we discuss the (local) behaviour of an objective function subject to constraints, and stability properties. This includes a brief account on critical and stationary points. Now, suppose that a parametric optimization problem $P(t)$, $t \in \mathbb{R}$, is given. As the parameter t increases, quite complicated degenerate situations might occur. However, it is possible to perturb $P(\cdot)$ slightly into the parameteric problem $\tilde{P}(\cdot)$ in order that for $\tilde{P}(t)$ and increasing t, the types of degeneracies appearing can be described by means of (minimal and non-avoidable) elementary singularities. This will be the content of Section 2.5. Finally, in Section 2.6 we turn to a coarser framework for a parametric optimization problem $P(t)$, where the discussion is devoted to piecewise differentiable 'critical' curves resulting from an approach via piecewise differentiable mappings.

2.2 UNCONSTRAINED OPTIMIZATION PROBLEMS

We start with the local behaviour of a fixed function $f \in C^2(\mathbb{R}^n, \mathbb{R})$, and after that, we turn to a family of functions in which one real parameter is involved.

Let U, V be open subsets of $\mathbb{R}^n$, and $k \geqslant 1$ an integer. A bijective mapping $\Phi: U \to V$ with $\Phi \in C^k(U, V)$ and $\Phi^{-1} \in C^k(V, U)$ is called C^k diffeomorphism between U and V. Note that a C^k diffeomorphism can be regarded as a C^k change of coordinates. For a proof of Theorems 2.2.1 and 2.2.2 we refer to [120].

THEOREM 2.2.1   Let $f \in C^2(\mathbb{R}^n, \mathbb{R})$, $f(0) = 0$ and $Df(0) \neq 0$. Then there exist open neighbourhoods U, V of the origin and a C^2 diffeomorphism $\Phi: U \to V$ sending the origin onto itself, such that

$$f \circ \Phi^{-1}(y_1, \ldots, y_n) = y_1. \tag{2.2.1}$$

From Theorem 2.2.1 we see that a point $\bar{x} \in \mathbb{R}^n$ at which $Df(\bar{x}) \neq 0$ cannot be a candidate for a local minimizer for f. So, at a local minimizer $\bar{x}$ for f we necessarily have $Df(\bar{x}) = 0$. Points at which Df vanishes are called *critical points* for f. In particular, we see that a local minimizer is a critical point, but the converse is not true (e.g. consider $f(x) = -x^2$ at $x = 0$). A critical point $\bar{x} \in \mathbb{R}^n$ is said to be *non-degenerate* if $D^2 f(\bar{x})$ is non-singular, where the matrix $D^2 f(x)$ is the Hessian of f evaluated at x. The local behaviour of f at a non-degenerate critical point is completely determined by the number of negative eigenvalues of the Hessian $D^2 f(\bar{x})$, multiplicities being taken into account. This is the content of the following theorem, the so-called *Morse lemma*. Let QI (quadratic index) and QCI (quadratic co-index) denote the number of negative and positive eigenvalues of $D^2 f(\bar{x})$, respectively.

THEOREM 2.2.2   Let $f \in C^2(\mathbb{R}^n, \mathbb{R})$, $f(0) = 0$, $Df(0) = 0$ and $D^2 f(0)$ be non-singular. Let k be the quadratic index QI. Then, there exist open neighbourhoods U, V of the origin and a C^1 diffeomorphism $\Phi: U \to V$ sending the origin onto itself, such that

$$f \circ \Phi^{-1}(y_1, \ldots, y_n) = -y_1^2 - y_2^2 \cdots - y_k^2 + y_{k+1}^2 + \cdots + y_n^2. \tag{2.2.2}$$

For a proof of Theorem 2.2.2, which works for $f \in C^k(\mathbb{R}^n, \mathbb{R})$, $k \geqslant 3$, see also [165]. See Figures 2.1(a) and (b) for a picture clarifying Theorem 2.2.1 and Theorem 2.2.2, respectively.

Concerning non-degenerate critical points, a few additional important remarks are to be made. First of all, a non-degenerate critical point $\bar{x} \in \mathbb{R}^n$ is an isolated point within the set of all critical points. In fact, consider the C^1

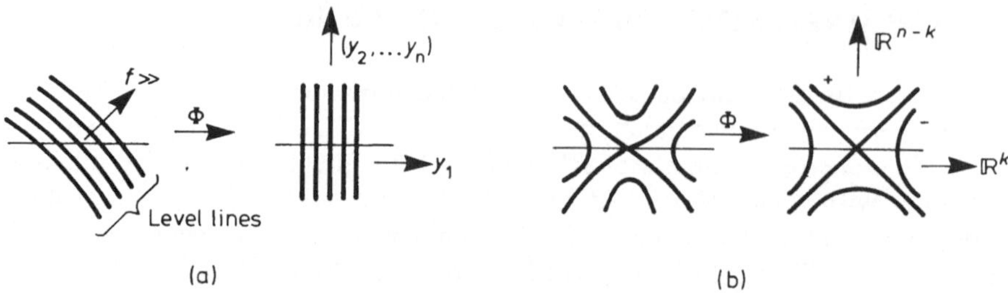

Figure 2.1

mapping $\mathcal{T}: x \mapsto D^{\mathrm{T}} f(x)$. Note that $\mathcal{T}(\bar{x}) = 0$ and that $D\mathcal{T}(\bar{x})(= D^2 f(\bar{x}))$ is non-singular. Hence, $\mathcal{T}$ is locally invertible by virtue of the inverse function theorem, and it follows that $\bar{x}$ is an isolated zero point for $\mathcal{T}$; so, $\bar{x}$ is an isolated critical point for f. By a similar argument we also obtain *stability* for the critical point (a precise definition of stability will be given in Section 2.4). In fact, let us embed the original function f into a parametric family $F(v, x)$. Here, v belongs to a Banach space V, $F(\bar{v}, \cdot) = f(\cdot)$ for some $\bar{v}$, and F is twice continuously Fréchet differentiable. Let $\bar{x}$ be a non-degenerate critical point for f, and consider the C^1 mapping $\mathcal{T}: (v, x) \mapsto D_x^{\mathrm{T}} f(v, x)$. It follows that $\mathcal{T}(\bar{v}, \bar{x}) = 0$ and $D_x \mathcal{T}(\bar{v}, \bar{x})$ is non-singular. But then we can apply the implicit function theorem (cf. [144]), thereby obtaining, locally, a unique C^1 function $x(v)$ with $x(\bar{v}) = \bar{x}$ and $\mathcal{T}(v, x(v)) \equiv 0$. This yields the C^1 depenence of the critical point $x(v)$ for $F(v, \cdot)$ on the parameter v. Now, by means of a suitable Banach space setting we can take the function f itself as a parameter (cf. [155], [121]), thereby obtaining the C^1 dependence of critical points on the problem data.

In spite of the fact that Theorems 2.2.1 and 2.2.2 are essentially of local nature, they can nevertheless be used in order to obtain an insight into the global behaviour of a function (also under constraints). To see this, let $f \in C^2(\mathbb{R}^n, \mathbb{R})$ be *non-degenerate* (i.e. all its critical points are non-degenerate). Then, up to constants, we can build up the function f by means of gluing 'elementary' functions of the type (2.2.1) and (2.2.2) together; see Figure 2.2. In this edification functions of the form (2.2.1) are only used to join the critical situations of the form (2.2.2). Then, there arises the question which combinations of local forms (2.2.2) can occur. The study of such combinations leads to a deformation of the continuous problem into an underlying combinatorial problem, and it forms the main body of the so-called Morse theory (cf. [165], [120], and for an intuitive introduction, [113]).

Now, a final problem is the description of the subset of $C^2(\mathbb{R}^n, \mathbb{R})$ consisting of non-degenerate functions. For this, we need the introduction of a topology on such a function space. For a good understanding of the topology used, note that $f \in C^2(\mathbb{R}^n, \mathbb{R})$ is non-degenerate if and only if $\Phi_f(x) > 0$ for all $x \in \mathbb{R}^n$,

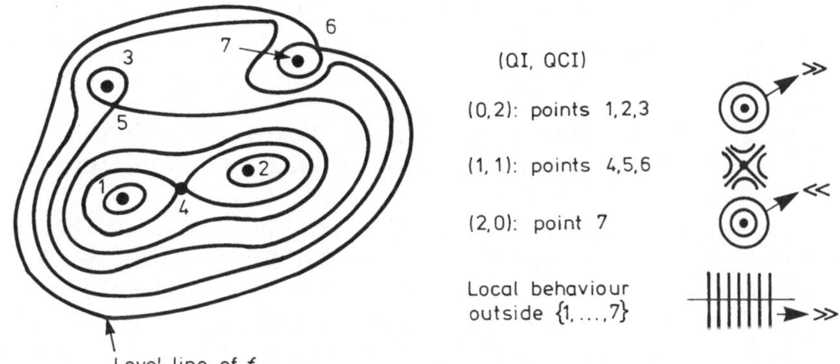

Figure 2.2

where

$$\Phi_f(x) = \| Df(x) \| + |\det D^2 f(x)|. \tag{2.2.3}$$

For a fixed non-negative integer k we introduce the C_s^k (or strong C^k) topology on $C^k(\mathbb{R}^n, \mathbb{R})$ by giving a basis for the topology (cf. [128]). Let $\alpha = (\alpha_1, \ldots, \alpha_n) \in \mathbb{N}^n$, $|\alpha| = \sum_{i=1}^n \alpha_i$, and denote by $\partial^\alpha f$ the αth partial derivative of f, where $f \in C^k(\mathbb{R}^n, \mathbb{R})$ and $|\alpha| \leqslant k$.

Put $C_+(\mathbb{R}^n, \mathbb{R}) = \{\Phi \colon \mathbb{R}^n \to \mathbb{R} \mid \Phi \text{ continuous and } \Phi(x) > 0 \text{ for all } x \in \mathbb{R}^n\}$.

DEFINITION 2.2.1 (cf. [121], [105]) For fixed $k \in \mathbb{N}$, a basis for the C_s^k topology for $C^k(\mathbb{R}^n, \mathbb{R})$ consists of all sets $V_{\Phi,f}^k$:

$$V_{\Phi,f}^k = \{g \in C^k(\mathbb{R}^n, \mathbb{R}) \mid |\partial^\alpha g(x) - \partial^\alpha f(x)| < \Phi(x), \text{ for all } x \in \mathbb{R}^n \text{ and all } \alpha \text{ with } |\alpha| \leqslant k\}. \tag{2.2.4}$$

The C_s^k topology for $C^k(\mathbb{R}^n, \mathbb{R}^m)$ will be the product C_s^k topology for $C^k(\mathbb{R}^n, \mathbb{R}) \times \cdots \times C^k(\mathbb{R}^n, \mathbb{R})$ (m times).

THEOREM 2.2.3 (cf. [121]) The subset $\mathscr{F}$ of $C^2(\mathbb{R}^n, \mathbb{R})$ consisting of non-degenerate functions is C_s^2 open and dense.

The open part of Theorem 2.2.3 is easily verified by taking the function Φ_f in (2.2.3) into account. The dense part relies on Sard's theorem on regular values.

DEFINITION 2.2.2 Let $F \in C^1(\mathbb{R}^n, \mathbb{R}^m)$. A point $\bar{x} \in \mathbb{R}^n$ is called a *critical* (*regular*) *point* for F if the induced linear map $\xi \mapsto DF(\bar{x})\xi$ is *not surjective* (*surjective*). A point $\bar{y} \in \mathbb{R}^m$ is called a *regular* (critical) *value* for F if $F^{-1}(\bar{y})$ contains no critical points (contains at least one critical point).

Theorem 2.2.4 (Sard's theorem, cf. [214]) Let $F \in C^k(\mathbb{R}^n, \mathbb{R}^m)$. If $k > \max(n - m, 0)$, then the set of *critical values* for F has the Lebesgue measure zero.

Let P be a property that holds on a set $A \subset \mathbb{R}^m$. Then P is said to hold for almost all points from $\mathbb{R}^m$ if the Lebesgue measure of $\mathbb{R}^m \backslash A$ vanishes. So, Sard's theorem reads: If $k > \max(n - m, 0)$, then almost all points in the target space are regular values for F. In particular, the set of regular values is dense.

Note that if $F^{-1}(\bar{y}) = \varnothing$, then $\bar{y}$ is automatically a regular value. In terms of regular valus we see: $f \in C^2(\mathbb{R}^n, \mathbb{R})$ is a non-degenerate function if and only if zero is a regular value for the 'critical point' map $x \mapsto D^{\mathrm{T}} f(x)$. As a consequence we obtain:

Corollary 2.2.5 (cf. [164]) Given $f \in C^2(\mathbb{R}^n, \mathbb{R})$. Then, the function $x \mapsto f(x) + v^{\mathrm{T}} x$ is non-degenerate for almost all $v \in \mathbb{R}^n$.

The above corollary can be used to prove the dense part of Theorem 2.2.3 by taking *local* non-degenerate approximations of f that have the form $f(x) + \xi(x) \cdot v^{\mathrm{T}} x$; here, ξ is a C^2 function vanishing outside some open neighbourhood of a given point $\bar{x}$ (cf. [121]).

Now we turn back to problems depending on a real parameter $t \in \mathbb{R}$. Let us write $\mathbb{R}^n \times \mathbb{R}$ for $\mathbb{R}^{n+1}$, and decompose a point $z \in \mathbb{R}^{n+1}$ as $z = (x, t)$, where $x \in \mathbb{R}^n$ and $t \in \mathbb{R}$.

Consider a function $f \in C^2(\mathbb{R}^n \times \mathbb{R}, \mathbb{R})$, $(x, t) \mapsto f(x, t)$. For each fixed $t \in \mathbb{R}$ we have a function of n variables, and we are interested in the set of critical points for $f(\cdot, t)$ depending on the parameter t. Thus, we define the critical point set Σ_{crit} as follows:

$$\Sigma_{\mathrm{crit}} = \{(x, t) \in \mathbb{R}^{n+1} \mid D_x f(x, t) = 0\}. \tag{2.2.5}$$

The closed set Σ_{crit} in (2.2.5) is determined by means of n equations in a space of $n + 1$ variables. Consequently, we generally expect Σ_{crit} to be a one-dimensional set. Indeed, this is true if Σ_{crit} can be described (locally) by using the implicit function theorem. For this, the Jacobian matrix $DD_x^{\mathrm{T}} f(z)$, being an $n \times (n + 1)$ matrix, should have rank n at all points z belonging to Σ_{crit}. However, the latter is true if and only if zero is a regular value for the mapping $(x, t) \mapsto D_x^{\mathrm{T}} f(x, t)$, and we obtain (cf. Figure 2.3):

Theorem 2.2.6 Let $f \in C^2(\mathbb{R}^n \times \mathbb{R}, \mathbb{R})$ and suppose that zero is a regular value for the mapping $\mathscr{T} : (x, t) \mapsto D_x^{\mathrm{T}} f(x, t)$. Then Σ_{crit} is a one-dimensional C^1 manifold

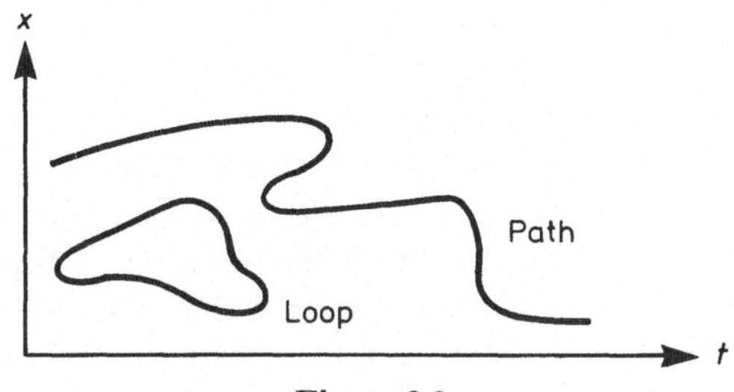

Figure 2.3

in $\mathbb{R}^n \times \mathbb{R}$. Moreover, each bounded connected component of Σ_{crit} is homeomorphic to a circle ('loop') and each unbounded connected component of Σ_{crit} is homeomorphic with $\mathbb{R}$ ('path').

Concerning Theorem 2.2.6 we note that a subset $M \subset \mathbb{R}^k$ is called a C^r manifold, $r \geqslant 1$, of dimension m if there exist, for each $\bar{x} \in M$, open neighbourhoods $\mathcal{O}$ and $\mathcal{V}$ of $\bar{x}$ and the origin, respectively, and a C^r diffeomorphism $\Phi \colon \mathcal{O} \to \mathcal{V}$ sending $\bar{x}$ onto the origin such that

$$\Phi(M \cap \mathcal{O}) = (\{0_{n-m}\} \times \mathbb{R}^m) \cap \mathcal{V}. \tag{2.2.6}$$

Furthermore, two subsets $A, B \subset \mathbb{R}^k$ are said to be homeomorphic if there exists a bijective mapping $\Phi \colon A \to B$ with both Φ and Φ^{-1} continuous (Φ is called a homeomorphism).

The last statement of Theorem 2.2.6 becomes obvious by means of a parametrization of Σ_{crit} by arc length.

As a corollary of Sard's theorem we obtain (note that f is to be taken as a C^3 function, now):

COROLLARY 2.2.7 Let $f \in C^3(\mathbb{R}^n \times \mathbb{R}, \mathbb{R})$. Then zero is a regular value for the mapping $(x, t) \mapsto D_x^{\mathsf{T}} f(x, t) + v$ for almost all v. In the case that zero is a regular value, the set Σ_{crit} belonging to the function $(x, t) \mapsto f(x, t) + v^{\mathsf{T}} x$ is a one-dimensional C^2 manifold.

Define

$$\mathscr{F}(A) = \{ f \in C^2(\mathbb{R}^n \times \mathbb{R}, \mathbb{R}) \mid \text{zero is a regular value for } (x, t) \mapsto D_x^{\mathsf{T}} f(x, t) \}.$$

Then, the following theorem can be obtained with the aid of Corollary 2.2.7 (cf. [121, Chapter 10]).

THEOREM 2.2.8 The set $\mathscr{F}(A)$ is C_s^2 open and dense in $C^2(\mathbb{R}^n \times \mathbb{R}, \mathbb{R})$.

Let $f \in \mathscr{F}(A)$. If $\bar{x} \in \mathbb{R}^n$ is a non-degenerate critical point for $f(\cdot, \bar{t})$, then, in a neighbourhood of $(\bar{x}, \bar{t})$, the set Σ_{crit} can be parametrized with the parameter t by means of the mapping $t \mapsto (x(t), t)$ obtained by application of the implicit function theorem. Then, by continuity, the number of negative (positive) eigenvalues of $D_x^2 f$ is constant at $(x(t), t)$ for t in a neighbourhood of $\bar{t}$. As a consequence of Theorem 2.2.2 the character of the critical point $x(t)$ for $f(\cdot, t)$ remains constant for t near $\bar{t}$. Next, suppose that x^i is a non-degenerate critical point for $f(\cdot, t^i)$, $i = 1, 2$, and let (x^1, t^1), (x^2, t^2) lie on the same component of Σ_{crit}. Then, walking from (x^1, t^1) to (x^2, t^2) along Σ_{crit}, a change of the character of the critical point for $f(\cdot, t)$ can only occur at points where $\det D_x^2 f$ vanishes. Since $f \in \mathscr{F}(A)$, we have rank $DD_x^T f = n$ along Σ_{crit} and, hence, rank $D_x^2 f \geqslant n - 1$. Put

$$\Sigma_{\text{deg}} = \{z \in \Sigma_{\text{crit}} \mid D_x^2 f(z) \text{ is singular}\}. \tag{2.2.7}$$

The latter rank observation implies that, passing a component of Σ_{deg} along Σ_{crit}, the quadratic index can change only by one. More precisely, in the latter situation the quadratic index does not change if and only if the restriction of the parameter t to Σ_{crit} is monotone; cf. Figure 2.4.

So far, we have not treated the parameter t as a special coordinate. In this way we discussed the critical set Σ_{crit} just as a one-dimensional curve in $\mathbb{R}^{n+1}$. This reflects a similar approach as in the later Section 2.6.

Now, we pay special attention to the parameter t, restricted to the set Σ_{crit}. In particular, we want to impose the property that $\det D_x^2 f$ changes the sign whenever passing a point from Σ_{deg} along Σ_{crit}. Regarding $\det D_x^2 f$ as a function on Σ_{crit}, a sufficient condition for changing the sign would be that the derivative of $\det D_x^2 f$ does not vanish whenever $\det D_x^2 f$ vanishes. In order to write down the latter condition, we need a higher degree of differentiability, and we assume $f \in C^3(\mathbb{R}^n, \mathbb{R})$. The condition we are looking for turns out to be the following extended regular value condition.

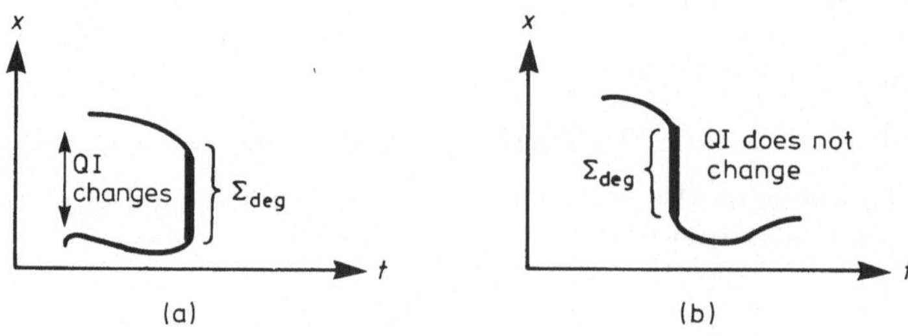

Figure 2.4

Condition B   Zero is a regular value for the mapping

$$\begin{pmatrix} x \\ t \end{pmatrix} \rightarrow \begin{pmatrix} D_x^{\mathrm{T}} f(x, t) \\ \det (D_x^2 f(x, t)) \end{pmatrix}.$$

Define

$$\mathscr{F}(B) = \{ f \in C^3(\mathbb{R}^n \times \mathbb{R}, \mathbb{R}) \mid f \text{ satisfies condition B} \}.$$

For a proof of the next theorem, compare [121, Chapter 10].

THEOREM 2.2.9

(i) The set $\mathscr{F}(B)$ is C_s^3 open and dense in $C^3(\mathbb{R}^n \times \mathbb{R}, \mathbb{R})$.

(ii) $\mathscr{F}(B) \subset \mathscr{F}(A)$.

(iii) An $f \in C^3(\mathbb{R}^n \times \mathbb{R}, \mathbb{R})$ belongs to $\mathscr{F}(B)$ if and only if $f \in \mathscr{F}(A)$, and, moreover, $\det D_x^2 f$, regarded as a function on Σ_{crit}, has a non-vanishing derivative at the zero points.

(iv) An $f \in C^3(\mathbb{R}^n \times \mathbb{R}, \mathbb{R})$ belongs to $\mathscr{F}(B)$ if and only if $f \in \mathscr{F}(A)$, and, moreover, if the parameter t, regarded as a function on Σ_{crit}, is non-degenerate.

Statement (iv) in Theorem 2.2.9 requires some explanation. First note that, if $f \in C^3(\mathbb{R}^n \times \mathbb{R}, \mathbb{R})$ belongs to $\mathscr{F}(A)$, then Σ_{crit} is a C^2 manifold. A C^2 function Φ on Σ_{crit} is called non-degenerate if it is non-degenerate (i.e. if it has only non-degenerate critical points) with respect to any local C^2 parametrization of Σ_{crit}. Note that, since Σ_{crit} is one-dimensional, non-degenerate critical points are either local minimizers or local maximizers. In particular, when taking the linear function $\Phi(x, t) \equiv t$, we see that the set Σ_{crit}, in a neighbourhood of a point from Σ_{deg}, can be approximated by means of a parabola ($f \in \mathscr{F}(B)$). In the latter case, the points from Σ_{deg} are called quadratic turning points and we emphasize that Σ_{deg} is a discrete subset of Σ_{crit}; see Figure 2.5(a). Note that, as t increases, at a quadratic turning point a pair of critical points is born, or a pair of critical points dies. In particular, the quadratic index changes exactly by one when passing a quadratic turning point along Σ_{crit}; see Figure 2.5(b).

A typical example of an $f \in \mathscr{F}(B)$ is the following one-dimensional one: $f(x, t) = \beta_1 x^3 + \beta_2 t x$, with $\beta_i \in \{ +1, -1 \}$, $i = 1, 2$. Note that the orientation of the parabola Σ_{crit} only depends on the product sign $\beta_1 \beta_2$. In particular, if $\beta_1 \beta_2 > 0$ (< 0), we are in the situation of Figure 2.5(b) I (II).

The above discussion on the one-dimensional case may be generalized as follows. Let $f \in \mathscr{F}(B)$ and $(\bar{x}, \bar{t}) \in \Sigma_{\mathrm{deg}}$. Let further $D_x^2 f(\bar{x}, \bar{t}) v = 0$, $v \neq 0$. So, v is an eigenvector to the (single) vanishing eigenvalue of $D_x^2 f(\bar{x}, \bar{t})$.

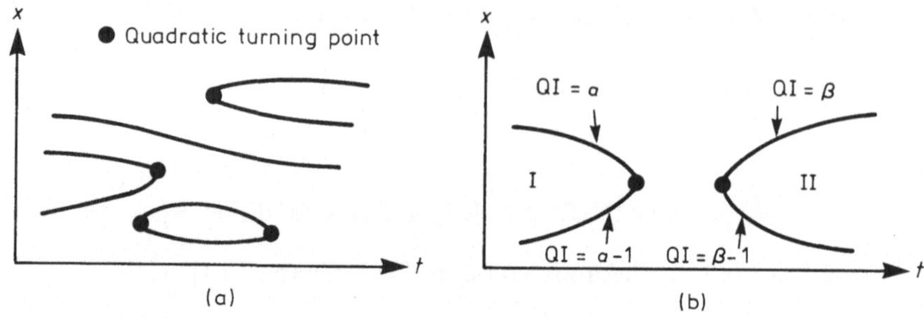

Figure 2.5

Put $\beta_1 = \operatorname{sign} D_x^3 f(\bar{x}, \bar{t})(v, v, v)$ and $\beta_2 = \operatorname{sign} D_t D_x f(\bar{x}, \bar{t})v$. Then, at $(\bar{x}, \bar{t})$, the set Σ_{crit} is oriented as in Figure 2.5(b) I (II) corresponding to $\beta_1 \beta_2 > 0$ (< 0); note that $\operatorname{sign}(\beta_1 \beta_2)$ does not depend on the choice of the eigenvector v.

The dense part of Theorem 2.2.9(i) can be achieved by means of (local) linear–quadratic perturbations of f. Let the space of real symmetric $n \times n$ matrices be identified with $\mathbb{R}^{(1/2)n(n+1)}$.

THEOREM 2.2.10   Let $f \in C^3(\mathbb{R}^n \times \mathbb{R}, \mathbb{R})$. Then, for almost all $(b, A) \in \mathbb{R}^n \times \mathbb{R}^{(1/2)n(n+1)}$, the function $(x, t) \mapsto f(x, t) + b^T x + \frac{1}{2} x^T A x$ belongs to $\mathscr{F}(B)$.

The discussion of the set Σ_{crit}, where the parameter t has a special meaning, reflects the study of constrained problems in Section 2.5.

The next theorem is an immediate consequence of the foregoing analysis (compare Figures 2.4 and 2.5(b); 'indoor–outdoor principle', 'corkscrewing' along Σ_{crit}.

THEOREM 2.2.11   Let $f \in \mathscr{F}(A)$. Suppose that $(x, 0)$, $(y, 1) \in \Sigma_{\text{crit}} \backslash \Sigma_{\text{deg}}$ lie on the same connected component of $\Sigma_{\text{crit}} \cap (\mathbb{R}^n \times [0, 1])$. Then, $\operatorname{QI}(x, 0) \equiv \operatorname{QI}(y, 1)$ (mod 2). Suppose that $(x, 0)$, $(y, 0) \in \Sigma_{\text{crit}} \backslash \Sigma_{\text{deg}}$, $x \neq y$, lie on the same connected component of $\Sigma_{\text{crit}} \cap (\mathbb{R}^n \times [0, 1])$. Then $\operatorname{QI}(x, 0) \equiv \operatorname{QI}(y, 0) + 1 \pmod{2}$.

2.3   CONSTRAINT SETS

We consider constraint sets (feasible sets) M of the following type, with $h_i, g_j \in C^1(\mathbb{R}^n, \mathbb{R})$:

$$M = \{x \in \mathbb{R}^n \mid h_i(x) = 0, \, i \in I, \, g_j(x) \leqslant 0, \, j \in J\}, \tag{2.3.1}$$

where

$$I = \{1,\ldots,m\}, \qquad m < n, \qquad J = \{1,\ldots,s\}. \qquad (2.3.2)$$

Define the activity map J_0 from $\mathbb{R}^n$ to the power set of J,

$$J_0(x) = \{j \in J \mid g_j(x) = 0\}. \qquad (2.3.3)$$

In the sequel we tacitly use the fact that $J_0(x) \subset J_0(\bar{x})$ for x near $\bar{x}$. The latter inclusion is implied by the continuity of the functions g_j and the fact that J is a finite set.

Unless additional specifications are made, the set M can be any closed subset of $\mathbb{R}^n$. Hence, in order to obtain a reasonable local structure for the set M, we have to impose additional assumptions, called constraint qualifications. In our study, two of them will play an important role:

(i) the linear independence constraint qualification (shortly LICQ);

(ii) the Mangasarian–Fromovitz constraint qualification (shortly MFCQ).

DEFINITION 2.3.1   At $\bar{x} \in M$ the *linear independence constraint qualification* is said to hold if the vectors $Dh_i(\bar{x})$, $i \in I$, $Dg_j(\bar{x})$, $j \in J_0(\bar{x})$, are linearly inependent.

DEFINITION 2.3.2   At $\bar{x} \in M$ the *Mangasarian–Fromovitz constraint qualification* is said to hold if the following two conditions are satisfied:

(MF1) The vectors $Dh_i(\bar{x})$, $i \in I$, are linearly independent,

(MF2) There exists a vector $\xi \in \mathbb{R}^n$ with

$$Dh_i(\bar{x})\xi = 0, \qquad i \in I, \qquad (2.3.4a)$$
$$Dg_j(\bar{x})\xi < 0, \qquad j \in J_0(\bar{x}). \qquad (2.3.4b)$$

The constraint qualifications LICQ is essentially stronger than MFCQ. If LICQ is satisfied at $\bar{x} \in M$, then, locally around $\bar{x}$, the set M has a nice structure in new differentiable coordinates. In fact, in this case, let us assume without loss of generality that $J_0(\bar{x}) = \{1,\ldots,p\}$. If $m + p < n$, choose vectors $\xi^{m+p+1},\ldots,\xi^n \in \mathbb{R}^n$, such that the set $\{D^{\mathsf{T}}h_i(\bar{x}),\ i \in I,\ D^{\mathsf{T}}g_j(\bar{x}),\ j \in J_0(\bar{x}),\ \xi^k,\ k \in \{m+p+1,\ldots,n\}\}$ forms a basis for $\mathbb{R}^n$. Next we put

$$
\begin{aligned}
y_i &= h_i(x), & i &= 1,\ldots,m \\
y_j &= -g_{j-m}(x), & j &= m+1,\ldots,m+p & &\text{shortly } y = \Phi(x). \quad (2.3.5)\\
y_k &= (x - \bar{x})^{\mathsf{T}}\xi^k & k &= m+p+1,\ldots,n
\end{aligned}
$$

Then, $\Phi(\bar{x}) = 0$, $D\Phi(\bar{x})$ is non-singular and hence Φ is locally invertible. From the special structure of Φ we see

$$\Phi(M \cap U) = (\{0_m\} \times \mathbb{H}^p \times \mathbb{R}^{n-m-p}) \cap V, \tag{2.3.6}$$

where U and V are some open neighbourhoods of $\bar{x}$ and 0, respectively, and

$$\mathbb{H}^p = \{(y_1, \ldots, y_p) \in \mathbb{R}^p \mid y_i \geqslant 0, i = 1, \ldots, p\}. \tag{2.3.7}$$

Let $T_{\bar{x}}M$ denote the tangent space of M at $\bar{x}$, i.e.

$$T_{\bar{x}}M = \bigcap_{i \in I} \operatorname{Ker} Dh_i(\bar{x}) \cap \bigcap_{j \in j_0(\bar{x})} \ker Dg_j(\bar{x}). \tag{2.3.8}$$

We note that the vectors $D\Phi^{-1}(0)e^k$, $k = m + p + 1, \ldots, n$, form a basis for $T_{\bar{x}}M$, where e^k stands for the kth unit vector from $\mathbb{R}^n$.

In the case when MFCQ is satisfied, the local structure of M is more complicated and, in general, it cannot be described in an easy way by means of a *differentiable* change of coordinates. However, a *continuous* change of coordintes will do.

THEOREM 2.3.1 (see [87]) Suppose that MFCQ is satisfied at all points of M. Then, M is a topological manifold (with boundary) of dimension $n - |I|$.

Concerning Theorem 2.3.1, a subset $X \subset \mathbb{R}^n$ is a *topological manifold (with boundary)* of dimension m if for each $\bar{x} \in X$ there exist open neighbourhoods $\mathcal{O}$ and $\mathcal{V}$ of $\bar{x}$ and the origin, respectively, and a homeomorphism $\Phi: \mathcal{O} \to \mathcal{V}$ sending $\bar{x}$ onto the origin, such that

$$\Phi(X \cap \mathcal{O}) = \begin{cases} \text{either } (\{0_{n-m}\} \times \mathbb{R}^m) \cap \mathcal{V}, & (2.3.9a) \\ \text{or} \quad (\{0_{n-m}\} \times \mathbb{H} \times \mathbb{R}^{m-1}) \cap \mathcal{V}. & (2.3.9b) \end{cases}$$

See Figures 2.6 (a) and (b) for an illustration concerning the mapping Φ from (2.3.6) and (2.3.9), respectively.

One of the major features of the constraint qualification MFCQ is its intimate relation with the stability of the feasible set. As an abbreviation we put

$$H = (h_1, \ldots, h_m)^{\mathrm{T}}, \qquad G = (g_1, \ldots, g_s)^{\mathrm{T}}. \tag{2.3.10}$$

The feasible set M generated by means of the functions h_i, g_j will be denoted by $M[H, G]$. In the next definition and theorem the functions h_i, g_j are supposed to be taken from $C^2(\mathbb{R}^n, \mathbb{R})$.

DEFINITION 2.3.3 The feasible set $M[H, G]$ is called C_s^2 stable if there exists a C_s^2 neighbourhood $\mathcal{O}$ of (H, G) such that for every $(\tilde{H}, \tilde{G}) \in \mathcal{O}$ the corresponding set $M[\tilde{H}, , \tilde{G}]$ is homeomorphic with $M[H, G]$.

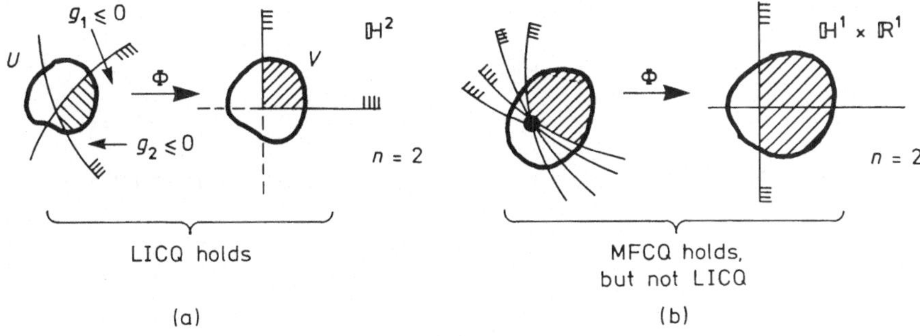

Figure 2.6

THEOREM 2.3.2 (see [87]) Let $M[H, G]$ be compact. Then, $M[H, G]$ is C_s^2 stable if and only if MFCQ is satisfied at all points $x \in M[H, G]$.

The latter theorem has an important impact on continuous deformations of optimization problems. Suppose that we have two optimization problems $P(0)$, $P(1)$, which are continuously connected by means of a parameter t, $t \in [0, 1]$. Suppose further that there is a compact set $K \subset \mathbb{R}^n$ containing the feasible set $M(t)$ belonging to problem $P(t)$ for all $t \in [0, 1]$. Now, if MFCQ is satisfied at every point of $M(t)$ for all $t \in [0, 1]$, then the final set $M(1)$ is homeomorphic with the starting set $M(0)$. This follows from Theorem 2.3.2 and the fact that the interval $[0, 1]$ is connected. In particular, if the topological structure of the set $M(1)$ differs from $M(0)$ (e.g. if $M(0)$ is connected, but $M(1)$ is not), then, for some $t \in [0, 1]$ at some $x \in M(t)$, the constraint qualification MFCQ must be violated. So, a violation of MFCQ is intimately related with bifurcational problems. For typical bifurcations appearing in one-parameter families $M(t)$ we refer to [121] and [115]. For stability results of the type of Theorem 2.3.2, which also take an objective function into account, we refer to [86].

2.4 CRITICAL POINTS, STATIONARY POINTS, STABILITY

In this section we deal with the constrained optimization problem (P),

$$(P) \qquad \text{Minimize } f(x) \text{ subject to } x \in M. \qquad (2.4.1)$$

The feasible set M is defined according to (2.3.1) and (2.3.2) by means of the constraint functions h_i, $i \in I$, and g_j, $j \in J$. The function f is called the objective function. All functions f, h_i, g_j are assumed to be taken from $C^2(\mathbb{R}^n, \mathbb{R})$.

DEFINITION 2.4.1 A point $\bar{x} \in M$ is called a *generalized critical point* (g.c. point) for (P) (or for $f|_M$) if the set of vectors

$$\{Df, Dh_i, Dg_j, i\in I, j\in J_0(\bar{x})\}|_{x=\bar{x}} \tag{2.4.2}$$

is linearly dependent.

So, when $\bar{x}\in M$ is a g.c. point, there exist λ, λ_i, $i\in I$, and μ_j, $j\in J_0(\bar{x})$, such that

$$\lambda Df + \sum_{i\in I}\lambda_i Dh_i + \sum_{j\in J_0(\bar{x})}\mu_j Dg_j|_{x=\bar{x}} = 0,$$

$$|\lambda| + \sum_{i\in I}|\lambda_i| + \sum_{j\in J_0(\bar{x})}|\mu_j| > 0. \tag{2.4.3}$$

Formally, we can define $\mu_j = 0$ for $j\notin J_0(\bar{x})$. Then, (2.4.3) can be altered by replacing the summation over $J_0(\bar{x})$ by a summation over J, thereby taking the following *complementarity condition* into account:

$$\sum_{j\in J}|\mu_j g_j(\bar{x})| = 0. \tag{2.4.4}$$

If $\bar{x}\in M$ is a local minimizer for $f|_M$, then $\bar{x}$ is a g.c. point and λ, μ_j in (2.4.3) can be chosen to be non-negative (cf. [104]). Such a g.c. point is also called a point of Fritz–John type. The validity of the constraint qualifications LICQ or MFCQ at a local minimizer $\bar{x}$ for $f|_M$ then implies that λ must be positive. In this case we can divide by λ and obtain in particular

$$Df + \sum_{i\in I}\lambda_i Dh_i + \sum_{j\in J}\mu_j g_j|_{x=\bar{x}} = 0,$$

$$\sum_{j\in J}|\mu_j g_j(\bar{x})| = 0. \tag{2.4.5}$$

DEFINITION 2.4.2   A g.c. point $\bar{x}\in M$ is called a *critical point* if both LICQ and relation (2.4.5) are satisfied. A g.c. point $\bar{x}\in M$ is called a *stationary point* if relation (2.4.5) holds with $\mu\geqslant 0$.

For $\bar{x}\in M$ any vector $(\bar{x},\lambda,\mu)$ satisfying (2.4.5) with $\mu\geqslant 0$ is called a Karush–Kuhn–Tucker point (KKT point).

The numbers λ_i, μ_j in (2.4.5) are called Lagrange parameters. If LICQ holds at $\bar{x}$, then, obviously, the Lagrange parameters are unique. Concerning MFCQ, we have the following compactness result of J. Gauvin. Put

$$\Lambda(\bar{x}) = \{(\lambda,\mu)\in\mathbb{R}^{|I|+|J|}|\mu\geqslant 0 \text{ and } (\lambda,\mu) \text{ satisfies } (2.4.5)\}. \tag{2.4.6}$$

THEOREM 2.4.1 (See [64]) Let $\bar{x}$ be a stationary point for (P). Then the set $\Lambda(\bar{x})$ is a compact convex polyhedron if and only if MFCQ is satisfied at $\bar{x}$.

We proceed with a discussion on *critical points*. So, let $\bar{x} \in M$ be a critical point for $f|_M$ with Lagrange parameters $(\bar{\lambda}, \bar{\mu})$. Without loss of generality we assume $J_0(\bar{x}) = \{1, \ldots, p\}$. We choose a local C^2 coordinate transformation Φ as defined in (2.3.5) and put

$$g(y) = f \circ \Phi^{-1}(y). \tag{2.4.7}$$

Note that the function g is nothing else but the function f in the new coordinates. In particular we have (cf. [120]).

$$\frac{\partial g}{\partial y_j}(0) = \bar{\mu}_{j-m}, \quad j = p+1, \ldots, p+m, \qquad \frac{\partial g}{\partial y_k}(0) = 0, \quad k = m+p+1, \ldots, n, \tag{2.4.8}$$

$$\frac{\partial^2 g}{\partial y_j \partial y_l}(0) = (e^k)^{\mathrm{T}}[D^{\mathrm{T}}\Phi^{-1}(0) \cdot D^2 L(\bar{x}) \cdot D\Phi^{-1}(0)]e^l, \quad k, l = m+p+1 \ldots, n, \tag{2.4.9}$$

where $e^k \in \mathbb{R}^n$ stands for the kth unit vector and L denotes the Lagrange function at $\bar{x}$:

$$L(x) = f(x) + \sum_{i \in I} \bar{\lambda}_i h_i(x) + \sum_{j \in J} \bar{\mu}_j g_j(x). \tag{2.4.10}$$

Recall that the feasible set M locally transforms into the set $\{0_m\} \times \mathbb{H}^p \times \mathbb{R}^{n-m-p}$ under Φ. So, by virtue of (2.4.8), the Lagrange parameters $\bar{\mu}_j$, $j \in J_0(\bar{x})$, just transform into first-order partial derivatives ($\mathbb{H}^p$). Moreover, in the directions $e^k, k = m+p+1, \ldots, n$, ($\mathbb{R}^{n-m-p}$) the first-order terms vanish, and second-order terms have to be taken into account there. Since the vectors $D\Phi^{-1}(0)e^k$, $k = m+p+1, \ldots, n$, form a basis for the tangent space $T_{\bar{x}}M$ (cf. Section 2.3), we see that the matrix

$$\left(\frac{\partial^2 g}{\partial x_k \partial x_l}(0)\right)_{k,l=m+p+1,\ldots,n}$$

represents the restriction of the Hessian of the Lagrangian $D^2 L(\bar{x})$ to the tangent space. The non-singularity of the latter restriction together with the non-vanishing of the Lagrange parameters $\bar{\mu}_j$, $j \in J_0(\bar{x})$, leads to the concept of a non-degenerate critical point. But, first, we explain the concept of 'restriction'.

Let A be a symmetric $n \times n$ matrix and $T \subset \mathbb{R}^n$ a linear subspace. A basis matrix V for T is a matrix with n rows whose columns form a basis for T. By the restriction of A to T, notation A/T, we mean some matrix of the family $\mathscr{V}$, $\mathscr{V} = \{V^{\mathrm{T}}AV \mid V$ a basis matrix for $T\}$. In view of Sylvester's theorem (cf. [155]), the number of negative, zero and positive eigenvalues of $V^T A V$, respectively, does not depend on the incidental choice of V. Therefore, the number of negative, zero and positive eigenvalues of A/T, respectively, is defined to be the corresponding number of $V^T A V$, where $V^T A V \in \mathscr{V}$. Furthermore, A/T is said to be non-singular if $V^T A V$ is non-singular, where $V^T A V \in \mathscr{V}$.

DEFINITION 2.4.3   Let $\bar{x} \in M$ be a critical point for $f|_M$ with Lagrange parameters $(\bar{\lambda}, \bar{\mu})$. The critical point $\bar{x}$ is called *non-degenerate* if the following two conditions are satisfied:

(ND1) $\bar{\mu}_j \neq 0$, $j \in J_0(\bar{x})$ ('strict complementarity'),

(ND2) $D^2 L(\bar{x})/T_{\bar{x}}M$ is non-singular.

The linear index LI (linear co-index LCI) is defined to be the number of negative (positive) numbers $\bar{\mu}_j$, $j \in J_0(\bar{x})$. The quadratic index QI (quadratic co-index QCI) is defined to be the number of negative (positive) eigenvalues of $D^2 L(\bar{x})/T_{\bar{x}}M$.

The local behaviour of $f|_M$ at a non-degenerate critical point $\bar{x}$ is completely determined by means of the indices (LI, LCI, QI, QCI), as expressed in the next theorem (compare Theorem 2.2.2 and see Figure 2.7). In particular, $\bar{x}$ is a local minimizer (maximizer) if and only if LI + QI = 0 (LCI + QCI = 0).

THEOREM 2.4.2   (See [120])   Let $\bar{x} \in M$ be a non-degenerate critical point for $f|_M$ with $f(\bar{x}) = 0$. Then, there exist open neighbourhoods U and V of $\bar{x}$ and the origin, respectively, and a C^1 diffeomorphism $\Phi: U \to V$ sending $\bar{x}$ to the origin, such that

$$\Phi(M \cap U) = (\{0_m\} \times \mathbb{H}^p \times \mathbb{R}^{n-m-p}) \cap V, \qquad (2.4.11)$$

and

$$f \circ \Phi^{-1}(0, \ldots, 0, y_{m+1}, \ldots, y_n) = -\sum_{i_1} y_{i_1} + \sum_{i_2} y_{i_2} - \sum_{i_3} y_{i_3}^2 + \sum_{i_4} y_{i_4}^2. \qquad (2.4.12)$$

The number of negative (positive) linear terms in (2.4.12) equals LI (LCI), and the number of negative (positive) squares equals QI (QCI); the coordnates y_{i_1}, y_{i_2} have to be taken non-negative ($\mathbb{H}^p$).

Analogously as we discussed in the unconstrained case (Section 2.2), for constrained problems, too, non-degenerate critical points are isolated critical points and depend differentiably on the problem data. The only crucial point is to establish a mapping $\mathscr{T}$ that turns critical points into zero points. In fact, let

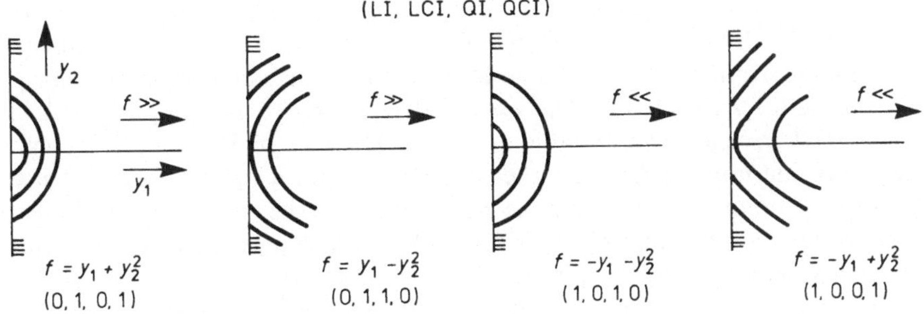

Figure 2.7

$\bar{x}$ be a critical point with Lagrange parameters (λ, μ). Define

$$\mathcal{F} : \begin{pmatrix} x \\ \lambda \\ \mu \end{pmatrix} \longmapsto \begin{pmatrix} D^{\mathrm{T}}f(x) + \sum_{i \in I} \lambda_i D^{\mathrm{T}}h_i(x) + \sum_{j \in J} \mu_j D^{\mathrm{T}}g_j(x) \\ h_i(x),\, i \in I \\ \mu_j g_j(x),\, j \in J \end{pmatrix}. \qquad (2.4.13)$$

Then, $\mathcal{F}$ is a C^1 mapping sending $(\bar{x}, \bar{\lambda}, \bar{\mu})$ onto zero. For the Jacobian matrix $D\mathcal{F}$ at $(\bar{x}, \bar{\lambda}, \bar{\mu})$ we obtain

$$D\mathcal{F}\begin{pmatrix} \bar{x} \\ \bar{\lambda} \\ \bar{\mu} \end{pmatrix} = \left(\begin{array}{c|c|c|c} D^2 L(x) & D^{\mathrm{T}}h_i(\bar{x}),\, i \in I & D^{\mathrm{T}}g_j(\bar{x}),\, j \in J_0(\bar{x}) & D^{\mathrm{T}}g_j(\bar{x}),\, j \notin J_0(\bar{x}) \\ \hline Dh_i(\bar{x}),\, i \in I & 0 & 0 & 0 \\ \hline \bar{\mu}_j Dg_j(\bar{x}),\, j \in J_0(\bar{x}) & 0 & 0 & 0 \\ \hline 0 & 0 & 0 & \mathrm{diag}(g_j(\bar{x}),\, j \notin J_0(\bar{x})) \end{array} \right).$$

$$(2.4.14)$$

From the next lemma it is easily seen that $D\mathcal{F}$ at $(\bar{x}, \bar{\lambda}, \bar{\mu})$ is non-singular if and only if $\bar{x}$ is a non-degenerate critical point.

LEMMA 2.4.3 (cf. [134]) Let A be a symmetric $n \times n$ matrix, B an $n \times k$ matrix and C a $k \times k$ matrix. Then, the matrix

$$\left(\begin{array}{c|c} A & B \\ \hline CB^{\mathrm{T}} & 0 \end{array} \right)$$

is non-singular if and only if C is non-singular, $\mathrm{rank}\,B = k$ and $A|_{\mathrm{Ker}\,B^{\mathrm{T}}}$ is non-singular.

So, if $\bar{x}$ is a non-degenerate critical point, then the mapping $\mathscr{T}$ is locally invertible and it follows that $\bar{x}$ is an isolated critical point. Now, we can put an additional parameter v (from some Banach space) into all functions f, h_i, g_j (and hence into $\mathscr{T}$) such that the extended mapping $(v, x, \lambda, \mu) \mapsto \mathscr{T}(v, x, \lambda, \mu)$ becomes a C^1 Fréchet differentiable mapping. Then, as in Section 2.2, we can make use of the implicit function theorem and obtain a local C^1 dependence of the critical point and its Lagrange parameters on the parameter v (cf. also [48]); in particular, the indices (LI, LCI, QI, QCI) and, hence, the local character of the critical point remain locally constant.

Let $H = (h_i)$ and $G = (g_j)$ be as in (2.3.10), and let $M[H, G]$ be the feasible set produced by H and G. Define $\mathscr{F}^*$ as follows:

$$\mathscr{F}^* = \{(f, H, G) \in C^2(\mathbb{R}^n, \mathbb{R})^{1+m+s} \mid M[H, G] \text{ satisfies LICQ at all of its points,}$$
$$\text{and all critical points for } f|_{M[H,G]} \text{ are nondegenerate}\}. \qquad (2.4.15)$$

Theorem 2.4.4 (cf. [121])   The set $\mathscr{F}^*$ is C_s^2 open and dense.

As we pointed out above, non-degenerate critical points (and their Lagrange parameters) depend C^1 on (C^2) perturbations of the problem data.

Another point of view, especially suitable for *stationary points* with (some) vanishing Lagrange parameters μ_j, $j \in J_0(\bar{x})$, is due to Kojima.

For $\alpha \in \mathbb{R}$, put

$$\alpha^+ = \max\{\alpha, 0\}, \qquad \alpha^- = \min\{\alpha, 0\}, \qquad (2.4.16)$$

and consider the mapping $\mathscr{H}$ (recall $I = \{1, \ldots, m\}$, $J = \{1, \ldots, s\}$):

$$\mathscr{H}: \begin{pmatrix} x \\ \lambda \\ \mu \end{pmatrix} \mapsto \begin{pmatrix} D^T f(x) + \sum_{i=1}^{m} \lambda_i D^T h_i(x) + \sum_{j=1}^{s} \mu_j^+ D^T g_j(x) \\ -h_i(x), \quad i = 1, \ldots, m \\ \mu_j^- - g_j(x), \, j = 1, \ldots, s \end{pmatrix}. \qquad (2.4.17)$$

The set of zero points of $\mathscr{H}$ and the set of Karush–Kuhn–Tucker points for (P) correspond to each other as follows. If (x, λ, μ) is a solution of $\mathscr{H} = 0$, then (x, λ, μ^+) is a KKT point, where $\mu^+ = (\mu_1^+, \ldots, \mu_s^+)$. Conversely, if (x, λ, μ) is a KKT point, then $(x, \lambda, \tilde{\mu})$ is a solution of $\mathscr{H} = 0$ with $\tilde{\mu}_j = \mu_j + g_j(x)$, $j \in J$. So, the mapping $\mathscr{H}$ in (2.4.17) is the right mapping when dealing with KKT points and, hence, with stationary points.

Unfortunately, due to the appearance of μ_j^+ and μ_j^-, the mapping $\mathscr{H}$ is not differentiable whenever μ_j vanishes. However, $\mathscr{H}$ is *piecewise continuously differentiable* (PC^1) with respect to the subdivision $\{\tau(\tilde{J}) \mid \tilde{J} \subset J\}$, where the cells $\tau(\tilde{J})$ are defined by

$$\tau(\tilde{J}) = \mathbb{R}^n \times \mathbb{R}^m \times \{\mu \in \mathbb{R}^s \mid \mu_j \geq 0, \, j \in \tilde{J}, \text{ and } \mu_j \leq 0, \, j \in J \setminus \tilde{J}\}, \qquad (2.4.18)$$

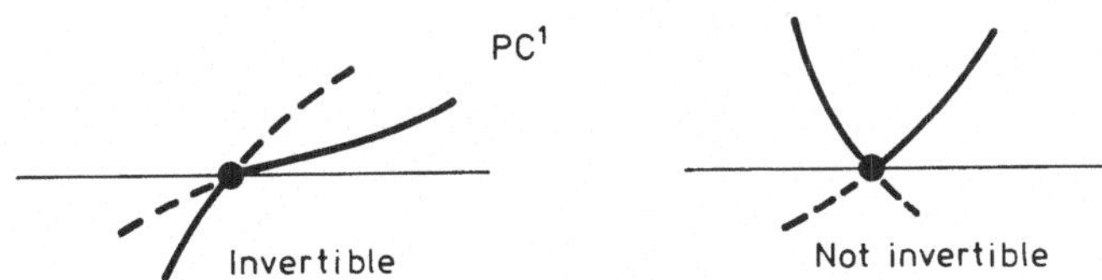

Figure 2.8

i.e. for each cell $\tau(\tilde{J})$ there exists an open neighbourhood U and a continuously differentiable mapping $\tilde{\mathscr{H}}: U \to \mathbb{R}^{n+m+s}$ with $\tilde{\mathscr{H}}|_{\tau(\tilde{J})} = \mathscr{H}|_{\tau(\tilde{J})}$.

Stability of stationary points can be studied with the aid of the mapping $\mathscr{H}$. In particular, if LICQ holds at $\bar{x}$, then stability is guaranteed if the Jacobian matrix $D\tilde{\mathscr{H}}$ of a C^1 extension of $\mathscr{H}$ to an open neighbourhood of $\tau(\tilde{J})$ at $(\bar{x}, \bar{\lambda}, \bar{\mu})$ is non-singular for all $\tilde{J} \subset J$ with common sign of the determinant (cf. [134]). This, in particular, guarantees the local invertibility of $\mathscr{H}$. For a simple clarification of the common sign condition, compare Figure 2.8.

Before stating conditions for local stability of stationary points in terms of derivatives of the underlying functions, we give a precise definition of this stability concept. Put again $H = (h_i)$ and $G = (g_j)$ as in (2.3.10), and let $M[H, G]$ denote the feasible set generated. For a given (f, H, G) and a subset $U \subset \mathbb{R}^n$ we put

$$\text{norm}[(f, H, G), U] = \sup_{x \in U} \max_{\Phi \in \{f, h_i, i \in I, g_j, j \in J\}} \left\{ |\Phi(x)| + \sum_i \left| \frac{\partial \Phi}{\partial x_i}(x) \right| + \sum_{i,j} \left| \frac{\partial^2 \Phi}{\partial x_i \partial x_j}(x) \right| \right\}.$$

(2.4.19)

For $x \in \mathbb{R}^n$ and $\delta > 0$ let $B(x, \delta)$ denote the Euclidean ball in $\mathbb{R}^n$, centred at x with radius δ.

DEFINITION 2.4.4 (See [134]) Let $x \in M[H, G]$ be a stationary point for $f|_{M[H,G]}$. Then, x is called *strongly stable* if for some $\bar{\delta} > 0$ and each $\delta \in (0, \bar{\delta}]$ there exists $\alpha > 0$ such that, whenever $(\tilde{f}, \tilde{H}, \tilde{G})$ satisfies $\text{norm}[(f - \tilde{f}, H - \tilde{H}, G - \tilde{G}), B(x, \bar{\delta})] \leqslant \alpha$, the ball $B(x, \delta)$ contains a stationary point that is unique in $B(x, \bar{\delta})$.

Note that strong stability in particular implies the continuous dependence of a stationary point on (C^2 perturbations of) the problem data.

For $\bar{x} \in \mathbb{R}^n$ and $\tilde{J} \subset J$ we put

$$T(\bar{x}, \tilde{J}) = \bigcap_{i \in I} \text{Ker } Dh_i(\bar{x}) \cap \bigcap_{j \in \tilde{J}} \text{Ker } Dg_j(\bar{x}). \qquad (2.4.20)$$

Moreover, given (λ, μ) we define

$$L_{[\lambda,\mu]}(x) = f(x) + \sum_{i\in I} \lambda_i h_i(x) + \sum_{j\in J} \mu_j g_j(x). \qquad (2.4.21)$$

THEOREM 2.4.5   (See [134])   Let $\bar{x}\in M$ be a stationary point for $f|_{M[H,G]}$.

(i) Let LICQ be satisfied at $\bar{x}$ and let (λ, μ) be the Lagrange parameters. Then, $\bar{x}$ is a strongly stable stationary point if and only if the Hessian $D^2 L_{[\lambda,\mu]}(\bar{x})$ has non-vanishing determinants with a common sign on the subspaces $T(\bar{x}, \tilde{J})$ with $J_+(\bar{x}) \subset \tilde{J} \subset J_0(\bar{x})$, where

$$J_+(\bar{x}) = \{\, j\in J_0(\bar{x})\,|\,\mu_j > 0\}. \qquad (2.4.22)$$

(ii) Let MFCQ be satisfied at $\bar{x}$, but not LICQ. Then, $\bar{x}$ is a strongly stable stationary point if and only if for every $(\lambda,\mu)\in\Lambda(\bar{x})$ the Hessian $D^2 L_{[\lambda,\mu]}(\bar{x})$ is positive definite on the subspace $T(\bar{x}, J_+(\bar{x}))$, with $J_+(\bar{x})$ as in (2.4.22).

As an illustration of Theorem 2.4.5, we present some examples depending on a parameter t.

Example 2.4.1

Let $f(x_1, x_2) = -x_1^2 + x_2^2$, and $g(x_1, x_2, t) = -x_2 + t\,(\leqslant 0)$. Then, the origin $0\in\mathbb{R}^2$ is a strongly stable stationary point for $t = 0$ (LICQ is satisfied); see Figure 2.9.

Example 2.4.2

Let $f(x_1, x_2) = x_1^2 - x_2^2$, and $g(x_1, x_2, t) = -x_2 + t\,(\leqslant 0)$. The origin $0\in\mathbb{R}^2$ is not srongly stable for $t = 0$ (LICQ is satisfied). See Figure 2.10 and note that there

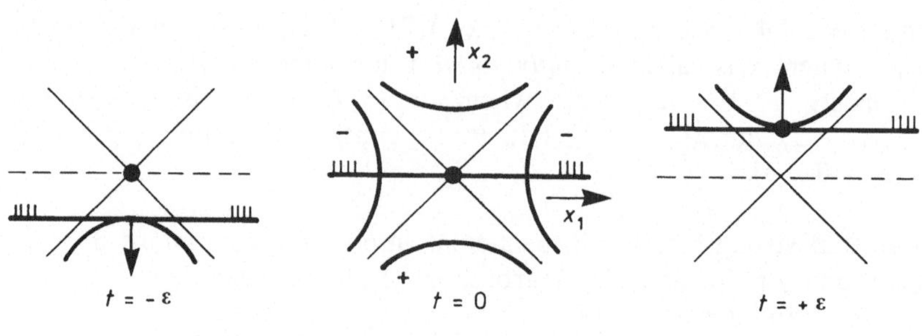

● Stationary point $(\varepsilon > 0)$

Figure 2.9

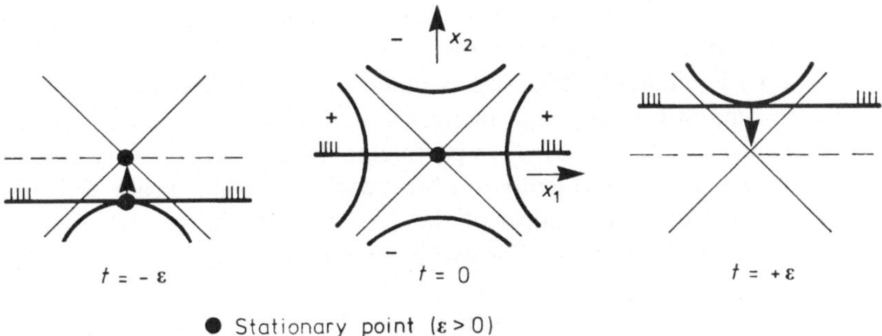

● Stationary point $(\varepsilon > 0)$

Figure 2.10

are two stationary points for $t < 0$, whereas there is no stationary point at all for $t > 0$.

Example 2.4.3

Let $f(x_1, x_2) = -x_1^2 + x_2^2$, and $g_1(x_1, x_2, t) = -x_2 + tx_1 + t$, $g_2(x_1, x_2, t) = -x_2 - tx_1 + t$. The origin $0 \in \mathbb{R}^2$ is not strongly stable (MFCQ is satisfied, but not LICQ); see Figure 2.11. Note that there are three stationary points for $t > 0$; compare also with Example 2.4.1!

We finish this section with some additional explanatory remarks, which also show the interaction of the concepts of 'non-degenerate critical point' and 'strongly stable stationary point'.

First, let *LICQ be satisfied at* $\bar{x} \in M$. Suppose that $\bar{x}$ is a strongly stable stationary point and recall statement (i) in theorem 2.4.5.

If $J_+(\bar{x}) = J_0(\bar{x})$, then $\bar{x}$ is a non-degenerate critical point (and, hence, stable via the implicit function theorem).

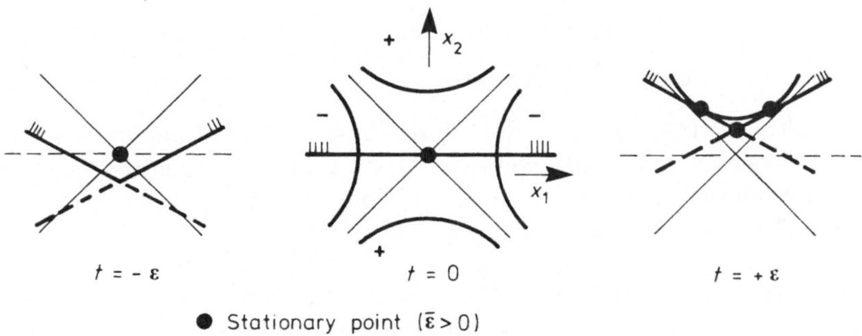

● Stationary point $(\bar{\varepsilon} > 0)$

Figure 2.11

Now suppose that $J_+(\bar{x}) \neq J_0(\bar{x})$; so there are some Lagrange parameters μ_j with $j \in J_0(\bar{x})$ that vanish (i.e. the strict complementarity condition (ND1) in Definition 2.4.3 is violated). The idea behind strong stability relies on the fact that the vanishing of Lagrange parameters μ_j ('linear terms') is compensated by means of 'positive quadratic terms'. In particular, if we enlarge the smallest linear subspace $T(\bar{x}, J_0(\bar{x}))$ monotonically until we reach the largest space $T(\bar{x}, J_+(\bar{x}))$, then the corresponding restriction of the Hessian $D^2_{[\lambda,\mu]}L(\bar{x})$ only enlarges its number of positive eigenvalues. This can be seen by means of the following two theorems from linear algebra.

Let A be a symmetric $n \times n$ matrix and $T \subset \mathbb{R}^n$ a linear subspace. Put $\pi(A/T) = (\rho, \eta, \theta)$, where ρ, η and θ are the number of positive, negative and zero eigenvalues of A/T, respectively, multiplicity being taken into account.

THEOREM 2.4.6 (See [98]) Let A be a symmetric $n \times n$ matrix and $T \subset \mathbb{R}^n$ a linear subspace. If A and A/T are non-singular, then

$$\pi(A) = \pi(A/T) + \pi(A^{-1}/T^{\perp}) \tag{2.4.23}$$

THEOREM 2.4.7 (see [41]) Let A be a non-singular symmetric $n \times n$ matrix and $T \subset \mathbb{R}^n$ a linear subspace. Then $A^{-1}/T^{\perp}$ is positive definite if and only if A has non-vanishing determinants with a common sign on the subspaces $\tilde{T}$ with $T \subset \tilde{T} \subset \mathbb{R}^n$.

From the discussion above we see that the number of negative eigenvalues of $D^2 L_{[\lambda,\mu]}(\bar{x})$ on $\tilde{T}$ is constant for all $T \subset \tilde{T} \subset \mathbb{R}^n$. This number is called the *stationary index* ([134]). In [87] a discussion of the stationary index in relation to the change of the topology of lower-level sets of the objective function at a strongly stable stationary point is presented.

For $\tilde{J}$ with $J_+(\bar{x}) \subset \tilde{J} \subset J_0(\bar{x})$ let $P(\tilde{J})$ be the following optimization problem

$$P(\tilde{J}): \qquad \text{minimize } f(x) \text{ subject to } x \in M(\tilde{J}), \tag{2.4.24}$$

$$M(\tilde{J}) = \{x \in \mathbb{R}^n \mid h_i(x) = 0, i \in I, g_j(x) = 0, j \in \tilde{J}\}. \tag{2.4.25}$$

Problem $P(\tilde{J})$ differs from problem (P) in the sense that the *inequality constraints* $g_j, j \in \tilde{J}$, are treated as *equality constraints*, and that the inequality constraints $g_j, j \notin \tilde{J}$, are omitted. Now, we see that the point $\bar{x}$ is a *non-degenerate* critical point for $P(\tilde{J})$ with *constant quadratic index* (in fact, equal to the stationary index). So, if we put an additional parameter v from some Banach space into all functions f, h_i, g_j such that the resulting functions become C^2 Fréchet differentiable, we obtain the parametric problems $P(v)$ and $P(\tilde{J})(v)$. By means of the implicit function theorem, the critical point for $P(\tilde{J})(v)$ near $\bar{x}$, say $x(\tilde{J})(v)$, depends C^1 on v. Finally, the stationary point $x(v)$ for $P(v)$, near $\bar{x}$, must be one of the critical points $x(\tilde{J})(v)$, with $J_+(\bar{x}) \subset \tilde{J} \subset J_0(\bar{x})$. Since strongly stable stationary points depend continuously on (C^2 perturbations of) the problem data, we see that $v \mapsto x(v)$ is a *continuous selection* of the C^1 mappings $v \mapsto x(\tilde{J})(v)$. In particular, this also implies local Lipschitz continuity for the mapping $v \mapsto x(v)$ (cf. [41]).

Now, let *MFCQ be satisfied at* $\bar{x} \in M$, *but not LICQ.* Moreover let $\bar{x}$ be a strongly stable stationary point. From Theorem 2.4.1 we know that the set $\Lambda(\bar{x})$ of Lagrange parameters ($\mu \geq 0!$) is a compact polyhedron. A compact polyhedron is the convex hull of its extreme points; it is not difficult to verify that $(\lambda, \mu) \in \Lambda(\bar{x})$ is an extreme point if and only if the corresponding vectors. $Dh_i(\bar{x})$, $i \in I$, $Dg_j(\bar{x})$, $j \in J_+(\bar{x})$, are linearly independent. Take an extreme point $(\lambda, \mu) \in \Lambda(\bar{x})$. Let $P[\lambda, \mu]$ be the optimization problem differing from (P) only in the sense that the index set of inequality constraints is equal to $J_+(\bar{x})$. Then we see that $\bar{x}$ is a non-degenerate critical point for $P[\lambda, \mu]$ with $LI = QI = 0$. Thus, the point $\bar{x}$ is a (non-degenerate) local minimizer for $P[\lambda, \mu]$ and, consequently, $\bar{x}$ is a local minimizer for problem (P). The stationary index at a strongly stable stationary point at which MFCQ, but not LICQ, is satisfied, is defined to be zero ([134]).

2.5 GENERIC SINGULARITIES IN ONE-PARAMETRIC OPTIMIZATION PROBLEMS

In this section we consider constrained optimization problems depending on one real parameter t, denoted by $P(t)$,

$P(t)$: $\qquad$ Minimize $f(x, t)$ subject to $x \in M(t)$, $\qquad$ (2.5.1)

$$M(t) = \{x \in \mathbb{R}^n \mid h_i(x, t) = 0, i \in I, g_j(x, t) \leq 0, j \in J\}, \qquad (2.5.2)$$

with

$$I = \{1, \ldots, m\}, \qquad m < n, \qquad J = \{1, \ldots, s\}. \qquad (2.5.3)$$

Throughout this section the functions f, h_i, g_j are assumed to be taken from $C^3(\mathbb{R}^n \times \mathbb{R}, \mathbb{R})$. Basic references concerning this section are [118] and [116]. In view of Theorem 2.4.4, for a particular fixed parameter value $\bar{t}$, we can expect that the feasible set $M(\bar{t})$ satisfies the constraint qualification LICQ at all of its points and, moreover, that the critical points for $f(\cdot \bar{t})|_{M\bar{t})}$ are non-degenerate. However, as the parameter t varies, we cannot avoid that LICQ or the non-degeneracy of critical points breaks down at certain points x for certain parameter values t. Although degeneracies will occur, we can avoid certain highly degenerate cases by means of (C_s^3) perturbations of the functions f, h_i, g_j. Here we will discuss the set of generalized critical points, resp. stationary

points, from the latter point of view. In fact, it is possible to classify those types of g.c. points that will occur within precisely five types in general.
Define

$$\Sigma_{gc} = \{(x,t) \in \mathbb{R}^n \times \mathbb{R} \,|\, x \text{ is a g.c. point for problem } P(t)\}. \tag{2.5.4}$$

$$\Sigma_{stat} = \{(x,t) \in \mathbb{R}^n \times \mathbb{R} \,|\, x \text{ is a stationary point for problem } P(t)\}. \tag{2.5.5}$$

The set Σ_{gc} (Σ_{stat}) can be understood as an unfolding of a particular g.c. point (stationary point) for a particular parameter value. With Section 2.2 in mind, we can expect that the sets Σ_{gc} and Σ_{stat} are one-dimensional sets in general.

The 'five types' of g.c. points, which we referred to above, will be desccribed and explained later. In the next definition we take this description already for granted. Define

$$H = (h_1, \ldots, h_m)^{\mathsf{T}}, \qquad G = (g_1, \ldots, g_s)^{\mathsf{T}}, \tag{2.5.6}$$

$$\mathscr{F}^{**} = \{(f, H, G) \,|\, \text{Each point of } \Sigma_{gc} \text{ belongs to type } 1, 2, 3, 4, 5\}. \tag{2.5.7}$$

THEOREM 2.5.1   (see [118])   The set $\mathscr{F}^{**}$ is C_s^3 open and dense in $C^3(\mathbb{R}^n \times \mathbb{R}, \mathbb{R})^{1+m+s}$.

Let z denote the general point in $\mathbb{R}^n \times \mathbb{R}$, and put $z = (x,t)$, $x \in \mathbb{R}^n$, $t \in \mathbb{R}$. Define

$$\Sigma_{gc}^i = \{z \in \Sigma_{gc} \,|\, z \text{ is of type } i\}, \qquad i = 1, 2, 3, 4, 5. \tag{2.5.8}$$

THEOREM 2.5.2   (see [118])   Let (f, H, G) belong to $\mathscr{F}^{**}$. Then we have Σ_{gc}^1 is open and dense in Σ_{gc} and, for $i = 2, 3, 4, 5$, the set Σ_{gc}^i is a discrete point set.

We proceed with a description of the five types. For each one we start with conditions that are necessary in order to describe the type under consideration, and we introduce so-called *characteristic numbers*, which determine the essence of the type. On the basis of the latter information we will describe the local structure of the set Σ_{gc} and the index relations (LI, LCI, QI, QCI) involved.

2.5.1   Type 1

A generalized critical point $\bar{z} = (\bar{x}, \bar{t})$ is a point of type 1 if $\bar{z}$ is a non-degenerate critical point, i.e. $\bar{x}$ is a non-degenerate critical point for $P(\bar{t})$.

Characteristic numbers: LI, LCI, QI, QCI.

If $\bar{z} = (\bar{x}, \bar{t})$ is a point of type 1, then we can parametrize the set Σ_{gc} by means of the parameter t in a neighbourhood of $\bar{z}$, whereas the indices (LI, LCI, QI, QCI) remain constant; compare Section 2.4. Since f, h_i, g_j are C^3 functions, the set Σ_{gc} is a C^2 manifold around $\bar{z}$.

2.5.2 Type 2

A generalized critical point $\bar{z} = (\bar{x}, \bar{t})$ is a point of type 2 if the following conditions (A1)–(A6) are fulfilled:

(A1) $\bar{x}$ is a critical point for $P(\bar{t})$

(A2) $J_0(\bar{z}) \neq \varnothing$.

After renumbering we may assume that $J_0(\bar{z}) = \{1, \ldots, p\}$, $p \geqslant 1$. Then, we have (cf. (A1) and (2.4.5)):

$$D_x f + \sum_{i-1}^{m} \bar{\lambda}_i D_x h_i + \sum_{j=1}^{p} \bar{\mu}_j D_x g_j \big|_{z=\bar{z}} = 0. \tag{2.5.9}$$

(A3) In (2.5.9), exactly one of the Lagrange parameters $\bar{\mu}_j$ vanishes.

After renumbering we may assume that $\bar{\mu}_p - 0$ and $\bar{\mu}_j \neq 0$, $j = 1, \ldots, p-1$. Put

$$T = \bigcap_{i \in I} \operatorname{Ker} D_x h_i(\bar{z}) \cap \bigcap_{j \in J_0(\bar{z})} \operatorname{Ker} D_x g_j(\bar{z}), \tag{2.5.10}$$

$$\widetilde{T} = \bigcap_{i \in I} \operatorname{Ker} D_x h_i(\bar{z}) \cap \bigcap_{j \in J_0(\bar{z}) \setminus \{p\}} \operatorname{Ker} D_x g_j(\bar{z}), \tag{2.5.11}$$

$$L(z) = f(z) + \sum_{i=1}^{m} \bar{\lambda}_i h_i(z) + \sum_{j=1}^{p} \bar{\mu}_j g_j(\bar{z}). \tag{2.5.12}$$

(A4) $D_x^2 L(\bar{z})/T$ is non-singular

(A5) $D_x^2 L(\bar{z})/\widetilde{T}$ is non-singular.

Let B be an $n \times r$ matrix of rank r. By $B^\dagger$ we denote the matrix $(B^T B)^{-1} B^T$. In fact, $B^\dagger$ is the Moore–Penrose inverse of B.

Let W be a basis matrix for the linear space $\widetilde{T}$. Put $\psi = (h_1, \ldots, h_m, g_1, \ldots, g_{p-1})^T$ and define the $n \times 1$ vectors:

$$\alpha = -((D_x^T \psi)^\dagger)^T \cdot D_t \psi \big|_{z=\bar{z}}, \tag{2.5.13}$$

$$\beta = -W(W^T \cdot D_x^2 L \cdot W)^{-1} W^T (D_x^2 L \cdot \alpha + D_t D_x^T L) \big|_{z=\bar{z}}, \tag{2.5.14}$$

$$\gamma = D_x g_p(\bar{z})(\alpha + \beta) + D_t g_p(\bar{z}). \tag{2.5.15}$$

(A6) $\gamma \neq 0$.

Let δ_1 and δ_2 denote the number of negative eigenvalues of $D_x^2 L(\bar{z})/\widetilde{T}$ and

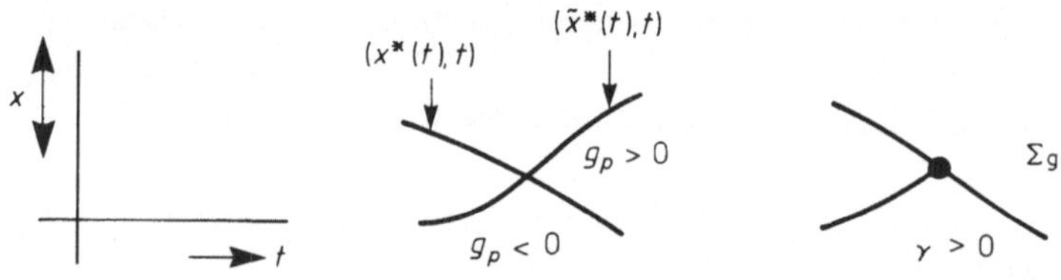

Figure 2.12

$D_x^2 L(\bar{z})/T$, respectively, and put $\delta = \delta_1 - \delta_2$.

Characteristic numbers: $\text{sign}(\gamma), \delta$.

A point of type 2 is a degenerate critical point; but, only the strict complementarity condition ((ND1) in Definition 2.4.3) is violated. Let $P^*(t)$ (resp. $\tilde{P}^*(t)$) denote the parametric optimization problem which differs from $P(t)$ in the sense that the *inequality constraint* g_p is turned into an *equality constraint* (resp. *omitted* as a constraint). Then $\bar{x}$ is a critical point for both $P^*(\bar{t})$ and $\tilde{P}^*(\bar{t})$; see condition (A3). Moreover, $\bar{x}$ is a non-degenerate critical point for both $P^*(\bar{t})$ and $\tilde{P}^*(\bar{t})$; compare conditions (A4) and (A5). As a consequence, the set Σ_{gc} is (locally) the union of the two curves $t \mapsto (x^*(t), t)$ and $t \mapsto (\tilde{x}^*(t), t)$ as far as they are feasible points; here, $x^*(t)$ and $\tilde{x}^*(t)$ are the critical points near $\bar{x}$ for $P^*(t)$ and $\tilde{P}^*(t)$, respectively. It can be shown that $d\tilde{x}^*(\bar{t})/dt = \alpha + \beta$ (cf. (2.5.13), (2.5.14)) and hence, if we follow the points $(\tilde{x}^*(t), t)$ for increasing t, we leave (enter) the feasible set $M(t)$ according to $\text{sign}(\gamma) = 1 \, (-1)$; see Figure 2.12. In Figure 2.13 all possible changes in the indices (LI, LCI, QI, QCI), determined by the characteristic numbers, are depicted.

2.5.3 Type 3

A generalized critical point $\bar{z} = (\bar{x}, \bar{t})$ is a point of type 3 if the following conditions (B1)–(B4) are fulfilled:

(B1) $\bar{x}$ is a critical point for $P(\bar{t})$.

In the case $J_0(\bar{z}) \neq \emptyset$, we may assume, after renumbering, that $J_0(\bar{z}) = \{1, \ldots, p\}$. From (B1) and (2.4.5) we see that the critical point relation (2.5.9) holds.

(B2) In (2.5.9) we have $\bar{\mu}_j \neq 0, j = 1, \ldots, p$.

Let the Lagrange function L be defined as in (2.5.12) and let the tangent space T be as in (2.5.10).

(B3) Exactly one eigenvalues of $D_x^2 L(\bar{z})/T$ vanishes.

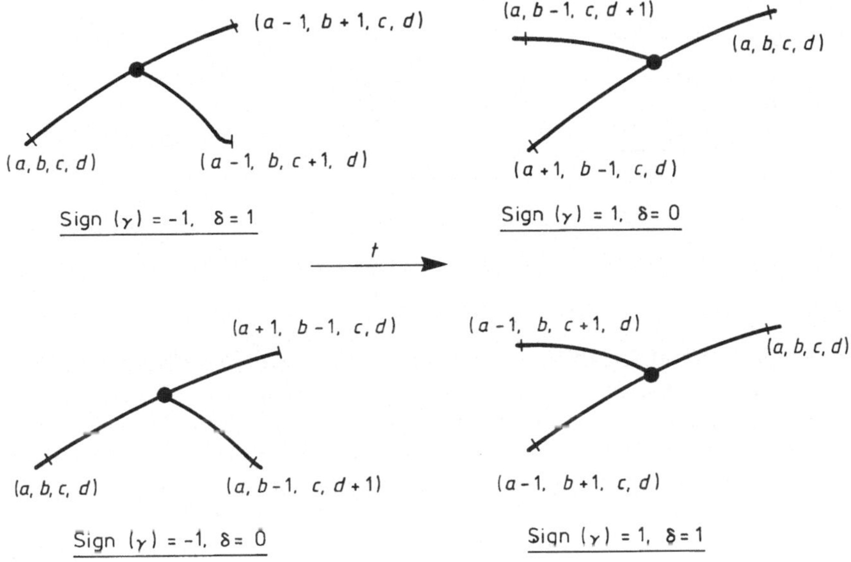

Figure 2.13

Let V be a basis matrix for the tangent space T. According to (B3), let w be a non-vanishing vector such that $V^T D_x^2 L(\bar{z}) V w = 0$, and put $v = V w$.

Put $\psi = (h_1, \ldots, h_m, g_1, \ldots, g_p)^T$ and define as in (2.5.13) (the symbol † denotes the Moore–Penrose inverse):

$$\beta_1 = D_x^3 L(v, v, v) - 3 v^T D_x^2 L \cdot ((D_x^T \psi)^\dagger)^T (v^T D_x^2 \psi v), \qquad (2.5.16)$$

$$\beta_2 = D_t(D_x L \cdot v) - D_t^T \psi \cdot (D_x^T \psi)^\dagger \cdot D_x^2 L v, \qquad (2.5.17)$$

where

$$v^T D_x^2 \psi v = (v^T D_x^2 h_1 v, \ldots, v^T D_x^2 g_p v)^T,$$

all partial derivatives being evaluated at $\bar{z}$. In the case that $I = J_0(\bar{z}) = \varnothing$, we have $T = \mathbb{R}^n$ and we omit all entries of ψ in (2.5.16) and (2.5.17). Next, we define

$$\beta = \beta_1 \cdot \beta_2. \qquad (2.5.18)$$

(B4) $\beta \neq 0$.

Let α denote the number of negative eigenvalues of $D_x^2 L(\bar{z})/T$.

Characteristic numbers: $\alpha, \operatorname{sign}(\beta)$.

A point of type 3 in a degenerate critical point; but, only the condition (ND2) in Definition 2.4.3 is violated. In fact, we deal with a direct generalization of the quadratic turning point as discussed in Section 2.2 around Figure 2.5.

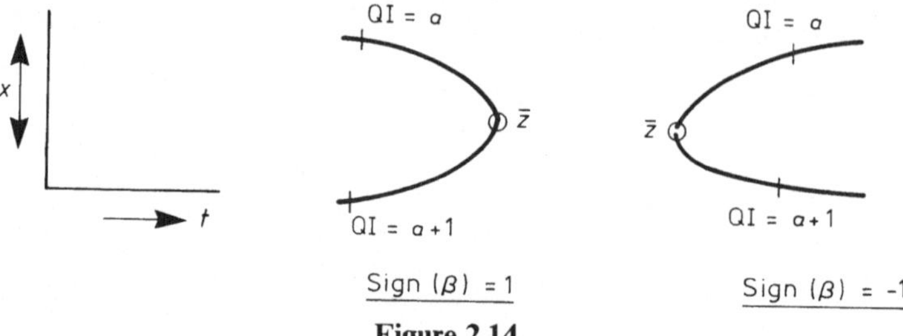

Figure 2.14

In a neighbourhood of $\bar{z}$ the index set of active inequality constraints for points on Σ_{gc} remains constant (hence, equal to $J_0(\bar{z})$). Locally around $\bar{z}$ the set Σ_{gc} is a one-dimensional C^2 manifold, and the linear function $\Phi(x,t)\equiv t$ has a non-degenerate local minimum (maximum) at $\bar{z}$ according to $\beta<0$ (>0). See Figure 2.14; since LI, LCI remain constant for points on Σ_{gc} near $\bar{z}$, only the change of QI is depicted.

2.5.4   Type 4

A generalized critical point $\bar{z} = (\bar{x}, \bar{t})$ is a point of type 4 if the following conditions (C1)–(C6) are fulfilled:

(C1) $|I| + |J_0(\bar{z})| > 0$.

In the case $J_0(\bar{z}) \neq \emptyset$, we may assume, after renumbering, that $J_0(\bar{z}) = \{1,\ldots,p\}$.

(C2) dim span $\{D_x h_i(\bar{z}), D_x g_j(\bar{z}), i\in I, j\in J_0(\bar{z})\} = m + p - 1$.

(C3) $m + p - 1 < n$.

From (C2) we see that there exist $\lambda_i, \mu_j, i\in I, j\in J_0(\bar{z})$, not all vanishing, such that

$$\sum_{i=1}^{m} \lambda_i D_x h_i(\bar{z}) + \sum_{j=1}^{p} \mu_j D_x g_j(\bar{z}) = 0. \tag{2.5.19}$$

Note that the numbers λ_i, μ_j in (2.5.19) are unique up to a common multiple.

(C4) In the case $p \neq 0$, we have $\mu_j \neq 0, j = 1,\ldots,p$, and we normalize the μ_j's by setting $\mu_p = 1$ (*normalization*).

Define

$$L(z) = \sum_{i=1}^{m} \lambda_i h_i(z) + \sum_{j=1}^{p} \mu_j g_j(z), \tag{2.5.20}$$

where λ_i, μ_j in (2.5.20) satisfy (2.5.19),

$$T = \bigcap_{i \in I} \operatorname{Ker} D_x h_i(\bar{z}) \cap \bigcap_{j \in J_0(\bar{z})} \operatorname{Ker} D_x g_j(\bar{z}). \tag{2.5.21}$$

Let W be a basis matrix for T. Define

$$A = D_t L \cdot W^T \cdot D_x^2 L \cdot W, \tag{2.5.22}$$

$$w = W^T \cdot D_x^T f, \tag{2.5.23}$$

all partial derivatives being evaluated at $\bar{z}$.

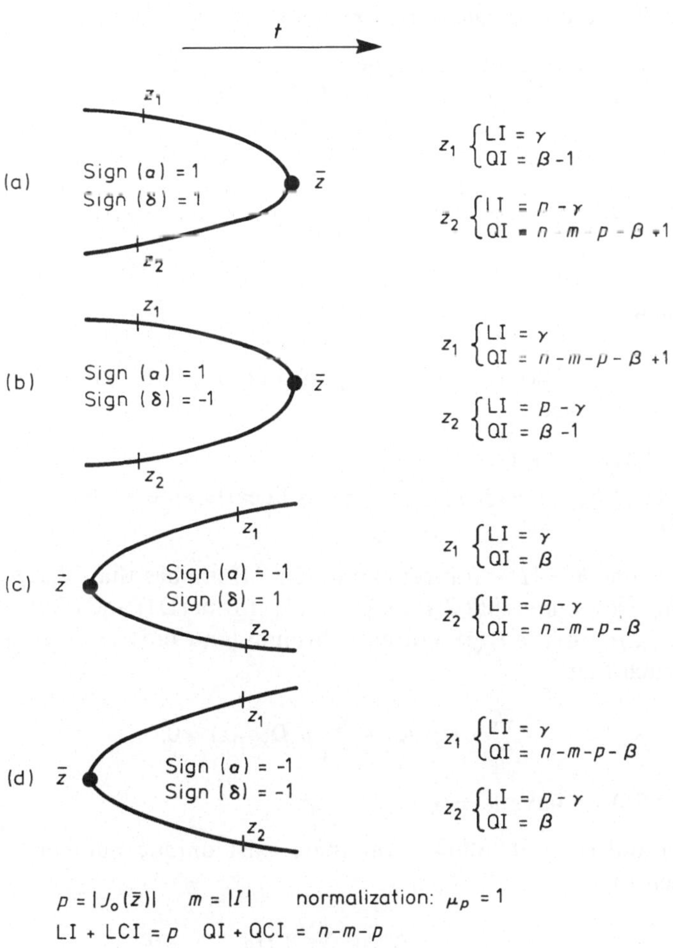

Figure 2.15

(C5) A is non-singular.

Finally, define

$$\alpha = w^{\mathrm{T}} A^{-1} w. \tag{2.5.24}$$

(C6) $\alpha \neq 0$.

We note that α is independent of the choice of the matrix W. Let β denote the number of *positive* eigenvalues of A. In the case $p \neq 0$, let γ be the number of negative $\mu_j, j \in \{1, \ldots, p-1\}$, and put $\delta = -D_t L(\bar{z})$. Note, in particular, that $\gamma < p$.

> *Characteristic numbers:* $\mathrm{sign}\,(\alpha), \beta$.

> *Characteristic numbers (corresponding to $\mu_p = 1$):* $\gamma, \mathrm{sign}(\delta)$.

For specific details about this type we refer to [118] and [116]. Here, we only mention that Σ_{gc}, in a neighbourhood of $\bar{z}$, is a one-dimensional C^2 manifold and the function $\Phi(x, t) \equiv t$ restricted to Σ_{gc} has a non-degenerate local minimum (maximum) at $\bar{z}$ corresponding to $\mathrm{sign}(\alpha) = -1\,(+1)$. So, in particular, Σ_{gc} has a quadratic turning point with respect to the parameter t at $\bar{z}$. For the behaviour of the indices (LI, LCI, QI, QCI) around $\bar{z}$, we refer to Figure 2.15.

2.5.5 Type 5

A generalized critical point $\bar{z} = (\bar{x}, \bar{t})$ is a point of type 5 if the following conditions (D1)–(D4) are fulfilled:

(D1) $|I| + |J_0(\bar{z})| = n + 1$. $\hspace{2cm}$ (2.5.25)

(D2) The set $\{Dh_i(\bar{z}), Dg_j(\bar{z}), i \in I, j \in J_0(\bar{z})\}$ is linearly independent (derivation in $\mathbb{R}^{n+1}$).

Since we assume $m = |I| < n$ throughout, (2.5.25) implies that $|J_0(\bar{z})| \geq 2$. After renumbering we assume that $J_0(\bar{z}) = \{1, \ldots, p\}$. From (D1) and (D2), we see that there exist $\lambda_i, \mu_j, i \in I, j \in J_0(\bar{z})$, not all vanishing (and unique up to a common multiple), such that

$$\sum_{i=1}^{m} \lambda_i D_x h_i(\bar{z}) + \sum_{j=1}^{p} \mu_j D_x g_j(\bar{z}) = 0. \tag{2.5.26}$$

(D3) In (2.5.26) we have $\mu_j \neq 0, j = 1, \ldots, p$.

From (D1) and (D2), it follows that there exist unique numbers $\alpha_i, \beta_j, i \in I, j \in J_0(\bar{z})$, such that

$$Df + \sum_{i=1}^{m} \alpha_i Dh_i + \sum_{j=1}^{p} \beta_j Dg_j\big|_{z=\bar{z}} = 0. \tag{2.5.27}$$

Put

$$\Delta_{ij} = \beta_i - \beta_j \cdot \mu_i / \mu_j, \qquad i,j = 1,\ldots,p, \tag{2.5.28}$$

and let Δ be a $p \times p$ matrix with Δ_{ij} as its (i,j)th element.

(D4) All off-diagonal elements of Δ are unequal to zero.

Put

$$L(z) = \sum_{i=1}^{m} \lambda_i h_i(z) + \sum_{j=1}^{p} \mu_j g_j(z), \tag{2.5.29}$$

where λ_i, μ_j satisfy (2.5.26). From (D2) we see that $D_t L(\bar{z}) \neq 0$. We define

$$\gamma_j = \text{sign}(\mu_j \cdot D_t L(\bar{z})), \qquad j = 1,\ldots,p. \tag{2.5.30}$$

Let δ_j denote the number of negative entries in the jth column of Δ, $j = 1,\ldots p$.

Characteristic numbers: $\gamma_j, \delta_j, j = 1,\ldots,p$.

For a local analysis of the set Σ_{gc} around $\bar{z}$ it is important to note, using (2.5.26) and (2.5.27), that

$$D_x f + \sum_{i=1}^{m} \left(\alpha_i - \beta_q \cdot \frac{\lambda_i}{\mu_q} \right) D_x h_i + \sum_{j=1}^{p} \left(\beta_j - \beta_q \cdot \frac{\mu_j}{\mu_q} \right) D_x g_j \big|_{z=\bar{z}} = 0 \qquad q = 1,\ldots,p.$$

$$\tag{2.5.31}$$

From the linear independence of $\{D_x h_i(\bar{z}), D_x g_j(\bar{z}), i \in I, j \in J_0(\bar{z}) \backslash \{q\}\}$, each $q \in \{1,\ldots,p\}$, and from (D4) we see, by virtue of (2.5.31), that $\bar{z}$ is a non-degenerate critical point if we omit g_p as a constraint. Put

$$M_q = \{z \mid h_i(z) = 0, g_j(z) = 0, i \in I, j \in J_0(\bar{z}) \backslash \{q\}\}, \qquad q = 1,\ldots,p, \tag{2.5.32}$$

$$M_q^- = \{x \in M_q \mid g_q(z) \leq 0\}. \tag{2.5.33}$$

From (D1), (D2) and the fact that h_i, g_j are C^3 functions it follows that, locally around $\bar{z}$, the set M_q is a one-dimensional C^3 manifold, $q = 1,\ldots,p$. Furthermore, in a neighbourhood of $\bar{z}$, the set Σ_{gc} is the union of the sets M_q^-, $q = 1,\ldots,p$. The indices (LI, LCI, QI, QCI) along $M_q^- \backslash \{\bar{z}\}$ are equal to $(\delta_q, p - 1 - \delta_q, 0, 0)$ with δ_q being one of the characteristic numbers (cf. (2.5.28) and (2.5.31)). Finally, as t increases and passes the value $\bar{t}$, the set M_q^- emanates from $\bar{z}$ (ends at $\bar{z}$) according to $\gamma_q = -1 (+1)$, where γ_q is a characteristic number.

Now we turn to a discussion of the set Σ_{stat} consisting of *stationary* points (cf. (2.5.5)). Let Σ_{reg} denote the set of non-degenerate critical points from Σ_{stat}, so (cf. (2.5.8))

$$\Sigma_{\text{reg}} = \{z \in \Sigma_{\text{gc}}^1 \mid \text{LI} = 0 \text{ at } z\}. \tag{2.5.34}$$

In particular we have the obvious inclusion

$$\Sigma_{\text{reg}} \subset \Sigma_{\text{stat}} \subset \bar{\Sigma}_{\text{reg}} \subset \Sigma_{\text{gc}}. \tag{2.5.35}$$

THEOREM 2.5.3   Let (f, H, G) belong to $\mathscr{F}^{**}$. Then $z \in \bar{\Sigma}_{\text{reg}} \setminus \Sigma_{\text{reg}}$ does not belong to Σ_{stat} if and only if z is of type 4.

Proof   It is easily seen that we only need to check the points of type 4, and in that case it suffices to show that $D_x f$ does not belong to the normal space $\mathscr{N} = \text{span}\{D_x h_i, i \in I, D_x g_j, j \in J_0(\bar{z})\}$. In fact, from (2.5.24) and (C6) we learn that the vector w does not vanish. If $D_x f$ belonged to $\mathscr{N}$, then we would have $D_x f \cdot \xi = 0$ for all $\xi \in T$, with T as in (2.5.11). But then, (2.5.23) would imply that w vanishes. This is a contradiction.

Let $\bar{z} \in \bar{\Sigma}_{\text{reg}} \setminus \Sigma_{\text{reg}}$ be of type 4. For $z \in \Sigma_{\text{reg}}$ and close to $\bar{z}$, let $\mu_j(z)$ denote the Lagrange parameter corresponding to the inequality constraint g_j, $j \in J_0(\bar{z})$. Then, $\mu_j(z)$ tends to infinity as z tends to $\bar{z}$, for all $j \in J_0(\bar{z})$. In the special case that $J_0(\bar{z}) = \varnothing$, some $\lambda_i(z)$ tends to infinity.

From Theorem 2.5.3 and (2.5.35) we see that the set Σ_{stat} coincides with $\bar{\Sigma}_{\text{reg}}$ *except* for the points of type 4. So, the local structue of Σ_{stat} is an immediate consequence of the local structure of the set $\bar{\Sigma}_{\text{reg}}$.

We say that MFCQ (resp. LICQ) holds at $z = (x, t)$ if MFCQ (resp. LICQ) holds at the point x with respect to problem $P(t)$.

THEOREM 2.5.4   (see [118])   Let (f, H, G) belong to $\mathscr{F}^{**}$, and let $\bar{z} \in \bar{\Sigma}_{\text{reg}}$. Then, the constraint qualification MFCQ is violated at $\bar{z}$ if and only if

either	$\bar{z}$ is of type 4
or	$\bar{z}$ is of type 5 and all μ_j in (2.5.26) have the same sign.

THEOREM 2.5.5   (see [118])   Let (f, H, G) belong to $\mathscr{F}^{**}$. Then, the set $\bar{\Sigma}_{\text{reg}}$ is a one-dimensional (piecewise C^2) manifold with boundary. In particular, $\bar{z} \in \bar{\Sigma}_{\text{reg}}$ is a boundary point if and only if we have at $\bar{z}$: $J_0(\bar{z}) \neq \varnothing$ and MFCQ is violated.

THEOREM 2.5.6   Let (f, H, G) belong to $\mathscr{F}^{**}$. Let Γ be a connected component of $\bar{\Sigma}_{\text{reg}}$. Then, Γ is homeomorphic to either a circle (loop), $\mathbb{R}^1$, $\mathbb{H}^1$, or the compact interval $[0, 1]$.

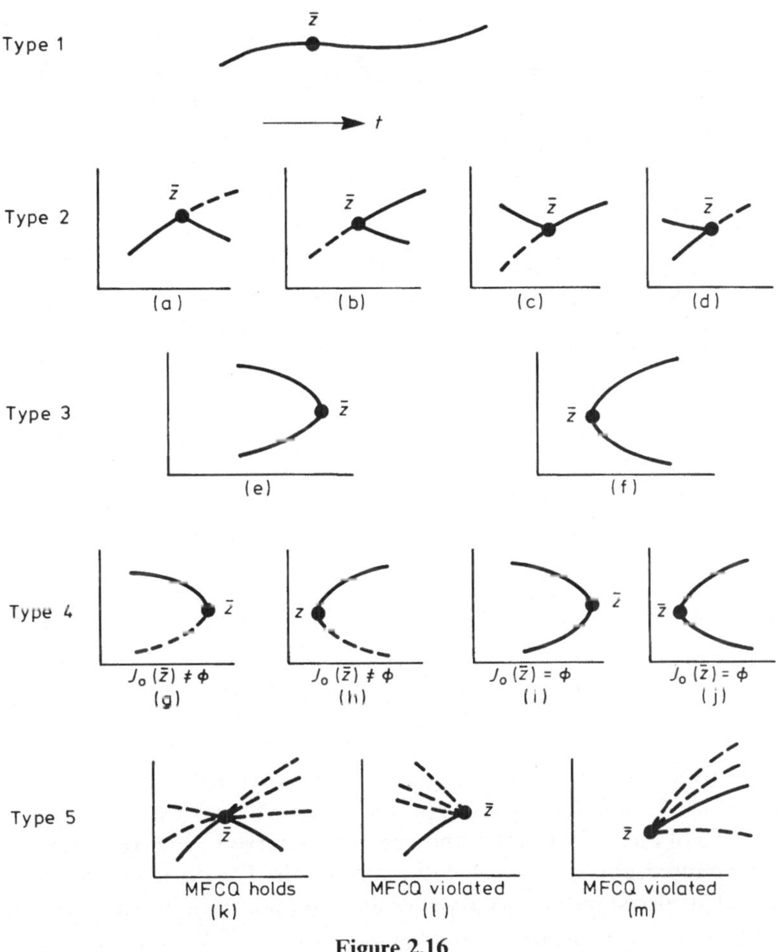

Figure 2.16

For a picture illustrating Theorem 2.5.5, see Figure 2.16. In Figure 2.16 the full curve stands for the curve of stationary points, and the dotted curve represents the curve of generalized critical points that are not stationary points.

Now we turn to a discussion of the set Σ_{loc} defined by

$$\Sigma_{loc} = \{(x, t) \in \mathbb{R}^n \times \mathbb{R} \mid x \text{ is a local minimizer for problem } P(t)\}. \qquad (2.5.36)$$

On the basis of the above investigation of Σ_{stat} we have the following possibilities for the local structure of $\bar{\Sigma}_{loc}$, with exactly one continuation in Σ_{loc}, Σ_{stat} and

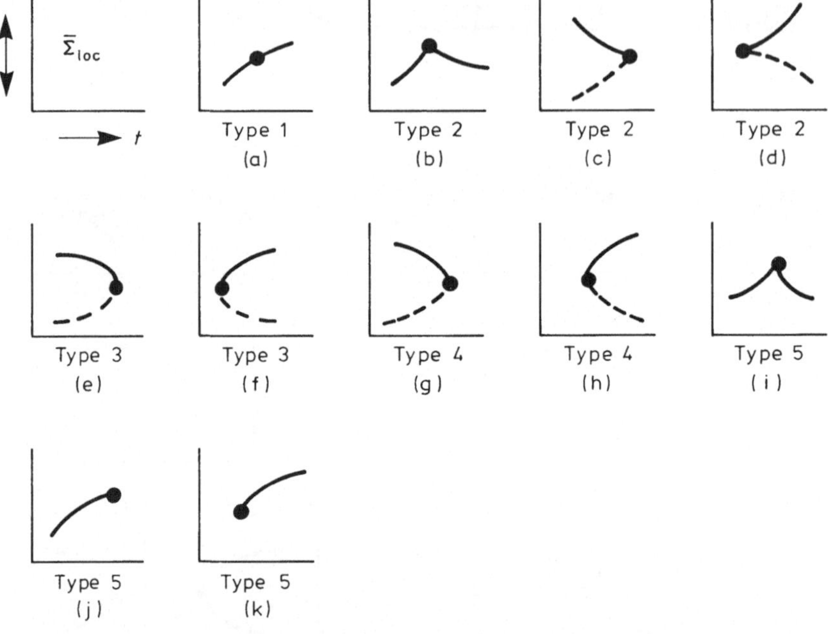

Figure 2.17

Σ_{gc}, respectively, as depicted in Figure 2.17. In this figure the point $\bar{z}$ under consideration is identified by an exposed point, whereas the full curve stands for the curve of local minimizers and the dotted curve in Figures 2.17(c), (d), (e) and (f) even represents a curve of stationary points. The dotted curve in Figures 2.17 (g) and (h) also stands for a curve of stationary points in the case $J_0(\bar{z}) = \varnothing$.

Finally, we present the appropriate generalization of Theorem 2.2.10. Let the space of symmetric $n \times n$ matrices (resp. $k \times l$ matrices) be identified with $\mathbb{R}^{(1/2)n(n+1)}$ (resp. $\mathbb{R}^{kl}$).

THEOREM 2.5.7   (see [199])   Let $(f, H, G) \in C^3(\mathbb{R}^n \times \mathbb{R}, \mathbb{R})^{1+m+s}$ be given. Then for almost all

$$(b, A, c, D, e, F) \in \mathbb{R}^n \times \mathbb{R}^{(1/2)n(n+1)} \times \mathbb{R}^m \times \mathbb{R}^{mn} \times \mathbb{R}^s \times \mathbb{R}^{sn}$$

the mapping

$$(x, t) \mapsto (f(x, t) + b^{\mathrm{T}}x + \tfrac{1}{2}x^{\mathrm{T}}Ax, H(x, t) + c + Dx, G(x, t) + e + Fx)$$

belongs to $\mathcal{F}^{**}$.

2.6 THE APPROACH VIA PIECEWISE DIFFERENTIABILITY

In this section we consider again the parametric optimization problem $P(t)$, $t \in \mathbb{R}$, as defined by (2.5.1)–(2.5.3); recall that $I = \{1, \ldots, m\}$, $m < n$, and $J = \{1, \ldots, s\}$. Unless stated otherwise, we assume that the functions f, h_i, g_j are taken from $C^2(\mathbb{R}^n \times \mathbb{R}, \mathbb{R})$. This setting up is due to Kojima [134], and Kojima and Hirabayashi [135], and we refer to these articles for specific details. We use the following parametric version of the mapping $\mathcal{H}$ from Section 2.4 (cf. (2.4.17)):

$$
\mathcal{H}: \begin{pmatrix} x \\ \lambda \\ \mu \\ t \end{pmatrix} \mapsto \begin{pmatrix} D_x^T f(x,t) + \sum_{i=1}^{m} \lambda_i D_x^T h_i(x,t) + \sum_{j=1}^{s} \mu_j^+ D_x^T g_j(x,t) \\ -h_i(x,t), i = 1, \ldots, m \\ \mu_j^- - g_j(x,t), j = 1, \ldots, s \end{pmatrix}. \tag{2.6.1}
$$

Let w denote the general point in $\mathbb{R}^{n+m+s+1}$, decomposed as

$$
w = (x, \lambda, \mu, t) \in \mathbb{R}^n \times \mathbb{R}^m \times \mathbb{R}^s \times \mathbb{R}. \tag{2.6.2}
$$

The mapping $\mathcal{H}$ is *piecewise continuously differentiable* (PC^1) with respect to the subdivision $\{\tau(\tilde{J}) \times \mathbb{R} \mid \tilde{J} \subset J\}$, where the cells $\tau(\tilde{J})$ are defined by (2.4.18); the factor $\mathbb{R}$ above merely represents the appearance of the parameter t. In spite of the fact that $\mathcal{H}$ is not differentiable, we can nevertheless define the concept of *regular value*. This will be done via the cell subdivision; we use the definition as formulated in [135].

DEFINITION 2.6.1 Zero is a regular value for $\mathcal{H}$ if for each $J_1 \subset J_2 \subset J$ there is an open set U containing $\sigma(J_2, J_1)$ and a continuously differentiable mapping $\tilde{\mathcal{H}}: U \to \mathbb{R}^{n+m+s}$ such that $\tilde{\mathcal{H}}|_{\sigma(J_2,J_1)} = \mathcal{H}|_{\sigma(J_2,J_1)}$ and the Jacobian matrix of $\tilde{\mathcal{H}} \circ i$ is of rank $n + m + s$ at each point v with $i(v) \in U$ and $\tilde{\mathcal{H}}(i(v)) = 0$, where

$$
\sigma(J_2, J_1) = \{(x, \lambda, \mu, t) \in \tau(J_2) \times \mathbb{R} \mid \mu_j = 0 \text{ for } j \in J_1\}, \tag{2.6.3}
$$

and where i is the natural embedding from $\mathbb{R}^n \times \mathbb{R}^m \times \mathbb{R}^{|J \setminus J_1|} \times \mathbb{R}$ into $\mathbb{R}^n \times \mathbb{R}^m \times \mathbb{R}^s \times \mathbb{R}$.

The set $\sigma(J_2, J_1)$ in (2.6.3) is a so-called face of $\tau(J_2) \times \mathbb{R}$. For the dimension we obviously have

$$
\dim \tau(J_2) \times \mathbb{R} = n + m + s + 1,
$$
$$
\dim \sigma(J_2, J_1) = \dim \tau(J_2) \times \mathbb{R} - |J_1|. \tag{2.6.4}
$$

Let zero be a regular value for $\mathcal{H}$. From (2.6.1) and (2.6.4) it follows that $\mathcal{H}$

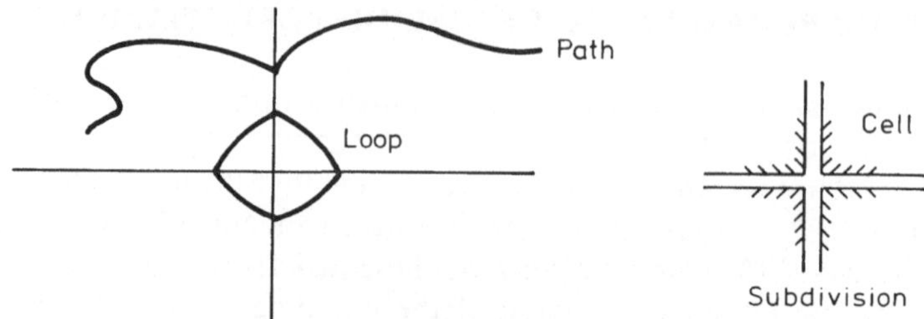

Figure 2.18

does not vanish on a face $\sigma(J_2, J_1)$ with $|J_1| > 1$. The set $\mathcal{H}^{-1}(0) \cap \sigma(J_2, \phi)$, if not empty, is a one-dimensional C^1 manifold by virtue of the implicit function theorem, meeting the faces $\sigma(J_2, J_1)$ with $|J_1| = 1$ transversally, i.e. non-tangentially. This gives rise to the following theorem.

THEOREM 2.6.1   (see [135])   Let zero be a regular value for the PC^1 mapping $\mathcal{H}$. Then, the set $\mathcal{H}^{-1}(0)$ is a one-dimensional (PC^1) manifold. Each bounded (resp. unbounded) connected component is homeomorphic with a circle (loop) (resp. the real line (path)).

See Figure 2.18 for an illustration.

Let Σ_{stat} again denote the set of stationary points. Then, Σ_{stat} is the image under the projection π of the set $\mathcal{H}^{-1}(0)$ to the (x, t)-space, where $\pi(x, \lambda, \mu, t) = (x, t)$. Under an additional assumption, the set Σ_{stat} is also a (PC^1) manifold.

Condition A   At every point of the *closure* of Σ_{stat} the constraint qualification MFCQ is satisfied.

THEOREM 2.6.2   (see [135])   Let zero be a regular value for $\mathcal{H}$ and suppose that condition A holds. Then, the set Σ_{stat} is a one-dimensional (PC^1) manifold. Each bounded (resp. unbounded) connected component is homeomorphic with a circle (loop) (resp. the real line (path)).

The underlying idea for the proof of Theorem 2.6.2 is the following. If, at a point $(\bar{x}, \bar{t}) \in \Sigma_{\text{stat}}$, the linear independence constraint qualification (LICQ) is violated, then (under the assumption of Theorem 2.6.2) the inverse image $\pi^{-1}(\bar{x}, \bar{t}) \subset \mathcal{H}^{-1}(0)$ is a line segment; see Figure 2.19.

Concerning the behaviour of the stationary index when passing points on Σ_{stat} that are not strongly stable, we present the next theorem.

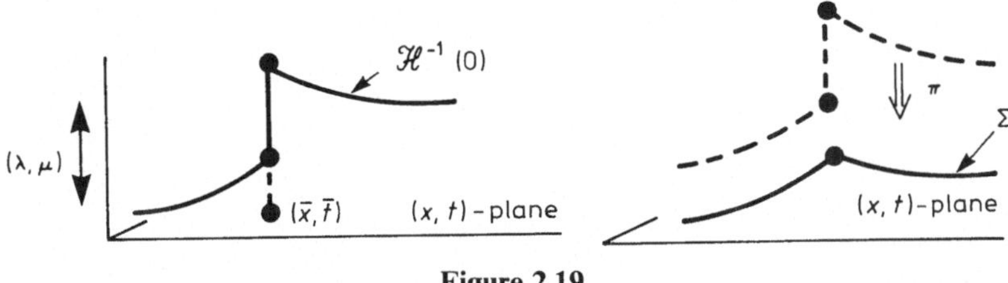

Figure 2.19

THEOREM 2.6.3 (see [135]) Suppose that zero is a regular value for $\mathcal{H}$ and that condition A holds. Let S be a connected component of Σ_{stat} and T a connected component of $S \cap \Sigma''$. Then, there exist an S neighbourhood V of T and a $p^* \in \{0, 1, \ldots, n - m - 1\}$ such that

$$V \cap \Sigma^s \subset \Sigma^s_{p^*} \subset \Sigma^s_{p^* - 1},$$

where

$$\Sigma^s = \{(x, t) \in \Sigma_{\text{stat}} \,|\, x \text{ is strongly stable for } P(t)\}, \qquad \Sigma'' - \Sigma_{\text{stat}} \backslash \Sigma^s, \qquad (2.6.5)$$

$$\Sigma^s_p = \{(x, t) \in \Sigma^s \,|\, \text{stationary index of } x \text{ equals } p\}. \qquad (2.6.6)$$

A theorem corresponding to Theorem 2.5.7 with H and G as in (2.5.6), cf. [135, Section 8], is the following one:

THEOREM 2.6.4 Let $(f, H, G) \in C^3(\mathbb{R}^n \times \mathbb{R}, \mathbb{R})^{1 + m + s}$ be given. Then, for almost all $(a, b, c) \in \mathbb{R}^n \times \mathbb{R}^m \times \mathbb{R}^s$, zero is a regular value for the PC^2 mapping $\mathcal{H}$ corresponding to the function

$$(x, t) \mapsto (f(x, t) + a^{\mathsf{T}}x, H(x, t) + b, G(x, t) + c).$$

In comparison to the set $\mathcal{F}^{**}$ introduced in Section 2.5, we note that $(f, H, G) \in \mathcal{F}^{**}$ implies that zero is a regular value for the associated mapping $\mathcal{H}$.

3

Pathfollowing of Curves of Local Minimizers

3.1 PRELIMINARY OUTLINE

We consider the following one-parametric optimization problem (cf. (1.1.1)):

$$P(t): \qquad \min \{ f(x,t) \mid x \in M(t) \}, \qquad t \in [t_A, t_B], \qquad (3.1.1)$$

where

$$M(t) := \{ x \in \mathbb{R}^n \mid h_i(x,t) = 0, i \in I, g_j(x,t) \leqslant 0, j \in J \}, \qquad (3.1.2)$$

with $I := \{1, \ldots, m\}$, $J := \{1, \ldots, s\}$, $m < n$, and $t_A < t_B$.

In this chapter we assume:

(E1) There exists a continuous function $x: [t_A, t_B] \to \mathbb{R}^n$ such that $x(t)$ is a local minimizer for $P(t)$.

(E2) $x(t_A)$ is known.

(V1) There exists a neighbourhood U of $\{(x(t), t) / t \in [t_A, t_B]\} \subset \mathbb{R}^n \times [t_A, t_B]$ such that for all $(x, t) \in U$ the functions f, g_i and h_j $(i = 1, \ldots, m; j = 1, \ldots, s)$ are twice continuously differentiable with respect to x.

(V2) The LICQ is satisfied at $x(t)$ for each $t \in [t_A, t_B]$ (cf. Definition 2.3.1).

Assumptions (V1) and (V2) imply that there exist functions $\lambda: [t_A, t_B] \to \mathbb{R}^m$, $\mu: [t_A, t_B] \to \mathbb{R}^s$, which are uniquely defined, such that $(x(t), \lambda(t), \mu(t))$ satisfies the KKT conditions (cf. Definition 2.4.2). Additionally, we need the following assumption (the so-called strong second-order sufficient condition):

(V3) $D_x^2 L(z(t)) \mid T_{x(t)}^+ M(t)$ is positive definite for all $t \in [t_A, t_B]$, where $z = (x, t)$ (in particular $z(t) = (x(t), t)$),

$$T^+_{x(t)}M(t):= \bigcap_{i\in I} \operatorname{Ker} D_x h_i(z(t)) \cap \bigcap_{j\in J^+(z(t))} \operatorname{Ker} d_x g_j(z(t)),$$

$$J^+(z(t)):= \{j\in J_0(z(t))|\mu_j(t) > 0\}$$

and

$$L(z) = f(z) + \sum_{i\in I} \lambda_i h_i(z) + \sum_{j\in J} \mu_j g_j(z).$$

It is easy to see that the assumptions (E1), (V1), (V2) and (V3) are satisfied if $(f, H, G)\in\mathscr{F}^{**}$ and $z(t)\in\Sigma^1_{\text{loc}} \cup \Sigma^2_{\text{loc}}$ for all $t\in[t_A, t_B]$, i.e. $x(t)$ is a local minimizer and a point of type 1 or type 2 (cf. Section 2.5).

From Section 2.5 we know that if we restrict ourselves to the class $\mathscr{F}^{**}$, then the assumptions (E1), (V2) and (V3) are not fulfilled for the interval $[0,1]$ in general. This is the reason why we consider an aribtrary interval $[t_A, t_B]$. With (E2) we have a starting point for the pathfollowing process. Section 3.2 includes an estimation of the radius of convergence of a general locally convergent algorithm (A) of the following structure. We determine a KKT point $w:=(v, t) = (x, \lambda, \mu, t)$ of $P(t)$ by means of an algorithm of the following kind:

(A) Start with v^0.

Having v^i, let v^{i+1} be a KKT point of the problem $P(v^i, t), i = 0, 1, 2, \ldots$, where

$$P(\bar{v}, t): \quad \min_x \{\varphi(x, \bar{v}, t)|g_j(\bar{x}, t) + D_x g_j(\bar{x}, t)(x - \bar{x})\leqslant 0, j\in J,$$

$$h_i(\bar{x}, t) + D_x h_i(\bar{x}, t)(x - \bar{x}) = 0, i\in I\}, \tag{3.1.3}$$

with

$$\varphi:\mathbb{R}^n \times \mathbb{R}^{n+m+s} \times [t_A, t_B] \to \mathbb{R}.$$

If there is more than one such point, choose v^{i+1} to be closest in norm to v^i.

Choosing special functions φ we obtain the following algorithms:

(1) Robinson's method ([187])

$$\varphi(x, \bar{v}, t):= f(x, t) + \sum_{i\in I} \bar{\lambda}_i(h_i(x, t) - h_i(\bar{x}, t) - D_x h_i(\bar{x}, t)(x - \bar{x}))$$

$$+ \sum_{j\in J} \bar{\mu}_j(g_j(x, t) - g_j(\bar{x}, t) - D_x g_j(\bar{x}, t)(x - \bar{x})).$$

(2) Wilson's method ([187], [229])

$$\varphi(x, \bar{v}, t):= D_x f(\bar{x}, t)(x - \bar{x}) + \tfrac{1}{2}(x - \bar{x})^T D_x^2 L(\bar{v}, t)(x - \bar{x}).$$

(3) Wilson's method with a consistent approximation for $D_x^2 L(.)$ ([61]):

$$\varphi(x, \bar{v}, t) := D_x f(\bar{x}, t)(x - \bar{x}) + \tfrac{1}{2}(x - \bar{x})^{\mathrm{T}} G(\bar{v}, t)(x - \bar{x}).$$

The jth column of $G(\bar{v}, t)$ is computed by

$$G(\bar{v}, t)_j := \frac{D_x^{\mathrm{T}} L(\bar{x} + \theta e_j, \bar{\lambda}, \bar{\mu}, t) - D_x^{\mathrm{T}} L(\bar{x}, \bar{\lambda}, \bar{\mu}, t)}{\theta} \qquad \text{for } \theta \neq 0,$$

where $\theta := r \, \| D_x L(\bar{v}, t) \|$ with some positive constant r, and

$$e^j := (0, \ldots, 1, \ldots, 0)^{\mathrm{T}} \qquad (j\text{th unit vector}).$$

Additionally we define

$$G(\bar{v}, t) := D_x^2 L(\bar{v}, t) \qquad \text{for } \theta = 0,$$

i.e. especially for $\bar{v} = v(t)$.

We observe the following common properties of these algorithms:

(Â1) If (V1) holds, then φ is twice continuously differentiable with respect to x for all $(x, \bar{v}, t) \in \mathbb{R}^n \times \mathbb{R}^{n+m+s} \times [t_A, t_B]$ with $(\bar{x}, t) \in U$ and $(x, t) \in U$.

(Â2) $D_x \varphi(\bar{x}, \bar{v}, t) = D_x f(\bar{x}, t)$ for all $\bar{v} \in \mathbb{R}^{n+m+s}$, for all $t \in [t_A, t_B]$.

(Â3) $D_x^2 \varphi(x(t), v(t), t) = D_x^2 L(v(t), t)$, for all $t \in [t_A, t_B]$.

The proof of Theorem 3.2.1 will be based on these common properties. The main tool for the proof is the implicit function theorem for generalized equations (cf. Robinson [190]). This investigation extends the results of Avila [11] to parametric optimization problems. In Section 3.2 we follow consistently the results given by Lehmann [140].

Based on the estimation of the rate of convergence of the locally convergent algorithm (A) on the interval $[t_A, t_B]$, a general pathfollowing procedure is proposed. We find a discretization

$$t_A = t_0 < \cdots < t_i < t_{i+1} < \cdots < t_N = t_B \tag{3.1.4}$$

and correspondingly sufficiently good approximations of $x(t_i)$, $i = 1, \ldots, N$, in a finite number of iteration steps. Of course, the discretization (3.1.4) can be computed as fine as necessary and the approximations of $x(t_i)$, $i = 1, \ldots, N$, as good as we want.

Section 3.3 contains an active index set strategy (cf. Gfrerer *et al.* [70] and Guddat *et al.* [93]), which reduces the computational expense since the size of the considered subproblems is smaller. In Section 3.4 the ALGORITHM PATH I and numerical results are presented.

3.2 THE ESTIMATION OF THE RADIUS OF CONVERGENCE

In this section we follow Lehmann (cf. [140]). We assume that (E1), (E2), (V1), (V2) and (V3) are fulfilled. Knowing the structure of (A) we will not only trace the path $\{(x(t), t) \mid t \in [t_A, t_B]\}$ as described in Section 3.1, but also the path of KKT points $\{(v(t), t) \mid t \in [t_A, t_B]\}$. Hence we will analyse the following radius of convergence:

DEFINITION 3.2.1 The radius of convergence of (A) solving $P(t)$ in a neighbourhood of $v(t)$ is defined by

$$r(t) := \sup \{r \mid \text{for all } v^0 \text{ with } \| v^0 - v(t) \| \leqslant r, \text{ the algorithm (A) starting with}$$
$$v^0 \text{ provides a sequence } \{v^i\},\ i = 1, 2, \ldots, \text{ converging to } v(t)\}. \quad (3.2.1)$$

The main result is contained in the following theorem:

THEOREM 3.2.1 (see [140]) Assume (E1), (V1), (V2) and (V3) for the optimization problems $P(t)$ and $(\hat{A}1)$, $(\hat{A}2)$ and $(\hat{A}3)$ for the locally convergent algorithm (A).

Let $r(t)$ be the radius of convergence of (A) solving $P(t)$ in a neighbourhood of $v(t)$.

Then there exists a real number $r > 0$ such that

$$r(t) \geqslant r \qquad \text{for all } t \in [t_A, t_B]. \qquad (3.2.2)$$

In preparation for the proof of Theorem 3.2.1, we introduce some lemmas:

LEMMA 3.2.1 Assume (E1), (V1) and (V2). Then the function $v \colon [t_A, t_B] \to \mathbb{R}^{n+m+s}$ defined by the KKT points of $P(t)$ is continuous.

Proof Because of (E1) we have to prove only the continuity of $(\lambda, \mu) \colon T \to \mathbb{R}^{m+s}$. Let $t_0 \in [t_A, t_B]$ and a sequence $\{t_k\}_{k=1,2,\ldots}$ converging to t_0 be arbitrarily chosen.

Case 1: $\{(\lambda(t_k), \mu(t_k))\}_{k=1,2,\ldots}$ is bounded. $(x(t_k), \lambda(t_k), \mu(t_k))$ fulfil the KKT condition

$$D_x L(x, \lambda, \mu, t) = 0, \qquad (3.2.3)$$

$$g_j(x, t) \leqslant 0, \qquad \mu_j \geqslant 0, \qquad \mu_j g_j(x, t) = 0, \qquad j = 1, \ldots, s, \quad (3.2.4)$$

$$h_i(x, t) = 0, \qquad i = 1, \ldots, m, \qquad (3.2.5)$$

for $t = t_k$ according to the definition. Because of (E1) and (V1) we obtain that

$(x(t_0), \lambda_0, \mu_0)$ satisfies (3.2.3)–(3.2.5) for any cluster point (λ_0, μ_0) of $\{(\lambda(t_k), \mu(t_k))\}$. Because of (V2) this system has a unique solution, hence $(\lambda_0, \mu_0) = (\lambda(t_0), \mu(t_0))$, which implies $(\lambda(t_k), \mu(t_k)) \to (\lambda(t_0), \mu(t_0))$ for $k \to \infty$.

Case 2: $\{(\lambda(t_k), \mu(t_k))\}_{k=1,2,\ldots}$ is unbounded. From (3.2.3) we obtain

$$D_x f(x(t_k), t_k) + \sum_{i \in I} \lambda_i(t_k) D_x h_i(x(t_k), t_k) + \sum_{j \in J} \mu_j(t_k) D_x g_j(x(t_k), t_k) = 0,$$

$$k = 1, 2, \ldots . \qquad (3.2.6)$$

As the sequence

$$\left\{ \frac{(\lambda(t_k), \mu(t_k))}{\| (\lambda(t_k), \mu(t_k)) \|} \right\}$$

($\| (\lambda(t_k), \mu(t_k)) \| \neq 0$ for k sufficiently large!) lies in the unit sphere, it has a cluster point $(\lambda^*, \mu^*) \neq 0$. After dividing (3.2.6) by $\| (\lambda(t_k), \mu(t_k)) \|$ we can conclude, regarding the continuity of $D_x f(.,.), D_x h_i(.,.), i \in I, D_x g_j(.,.), j \in J$, and $x(.)$ that

$$\sum_{i \in I} \lambda_i^* D_x h_i(x(t_0), t_0) + \sum_{j \in J} \mu_j^* D_x g_j(x(t_0), t_0) = 0. \qquad (3.2.7)$$

As $x(.), h_i(.,.), i \in I, g_j(.,.), j \in J$, are continuous, we obtain for k sufficiently large

$$g_j(x(t_k), t_k) < 0 \qquad \text{for } j \in J \setminus J_0(x(t_0), t_0)$$

and, because of (3.2.4),

$$\mu_j(t_k) = 0 \qquad \text{for } j \in J \setminus J_0(x(t_0), t_0).$$

Hence,

$$\mu_j^* = 0 \qquad \text{for } j \in J \setminus J_0(x(t_0), t_0). \qquad (3.2.8)$$

However, (3.2.7) and (3.2.8) together with $(\lambda^*, \mu^*) \neq 0$ contradict the linear independence of the gradients of the active constraints assumed by (V2). $\qquad \square$

The following lemma is a tool for the evaluations in the main proof:

LEMMA 3.2.2   Let the function $\Phi : \Omega \times P \subset \mathbb{R}^k \times \mathbb{R}^r \to \mathbb{R}^l$ be continuous and continuously differentiable with respect to the first variable in an open neighbourhood of $(x_0, p_0) \in \Omega \times P$. Then, for any $\mu > 0$ there are real numbers $R > 0$, $\rho > 0$ such that

$$\| \Phi(y, p) - \Phi(x, p) - D_x \Phi(x, p)(y - x) \| \leq \mu \| y - x \|$$

$$\forall x, y \in U_R(x_0), \quad \forall p \in U_\rho(p_0).$$

Proof   The proof consists of an application of [190], Theorem 2.2.8. $\qquad \square$

The proof of Theorem 3.2.1 will be based on Robinson's results about strongly regular generalized equations ([190]). First we introduce the meaning of this notation:

DEFINITION 3.2.2   Let Ω be an open subset of $\mathbb{R}^k$ containing a point x_0. Let C be a closed convex set in $\mathbb{R}^k$, and let $f:\Omega \to \mathbb{R}^k$ be differentiable at x_0. Suppose that the generalized equation

$$0 \in f(x) + \partial \Psi_c(x) \tag{3.2.9}$$

with

$$\partial \Psi_c(x) := \begin{cases} \{y \in \mathbb{R}^k \mid y^{\mathrm{T}}(c - x) \leqslant 0, \forall c \in C\}, & \text{if } x \in C, \\ 0, & \text{if } x \notin C, \end{cases} \tag{3.2.10}$$

has x_0 as a solution, and define, for $x \in \mathbb{R}^k$,

$$T(x) := f(x_0) + Df(x_0)(x - x_0) + \partial \Psi_c(x). \tag{3.2.11}$$

We say that (3.2.9) is strongly regular at x_0 (with the associated Lipschitz constant α) if there exist neighbourhoods U of the origin in $\mathbb{R}^k$ and V of x_0 such that the restriction to U of $T^{-1} \cap V$ is a single-valued function from U to V which is Lipschitzian on U with modulus α.

The following theorem represents a generalization of the well known implicit function theorem:

THEOREM 3.2.2   Let Ω, C, x_0 be as in Definition 3.2.2, $P \subseteq \mathbb{R}^r$, $p_0 \in P$, and let $F:\Omega \times P \to \mathbb{R}^k$. Suppose that F is continuous and continuously differentiable with respect to the first variable on $\Omega \times P$, and that x_0 solves

$$0 \in F(x, p_0) + \partial \Psi_c(x). \tag{3.2.12}$$

If (3.2.12) is strongly regular at x_0, with an associated Lipschitz constant α, then, for any $\varepsilon > 0$, there exist neighbourhoods N of p_0 and W of x_0, and single-valued function $x: N \to W$ such that for any $p \in N$, $x(p)$ is the unique solution in W of the inclusion

$$0 \in F(x, p) + \partial \Psi_c(x).$$

Further, for each p and q in N one has

$$\| x(p) - x(q) \| \leqslant (\alpha + \varepsilon) \| F(x(q), p) - F(x(q), q) \|.$$

Proof   See Robinson [190], Theorem 2.1. $\qquad\qquad\qquad\qquad\qquad\qquad\qquad$ $\square$

The following lemma shows that it suffices to prove only a 'local variant' of Theorem 3.2.1:

LEMMA 3.2.3   (cf. [11])   Let $r: [t_A, t_B] \to \mathbb{R}$ be a function with the following property: for each $t_0 \in [t_A, t_B]$ there exists a constant $\eta = \eta(t_0) > 0$ and a $\delta = \delta(t_0) > 0$ such that $r(t) \geq \eta$ for $|t - t_0| < \delta$.   (3.2.13)

Then there exists an $r > 0$ such that

$$r(t) \geq r \qquad \forall t \in [t_A, t_B].$$

Proof   The proof based on the compactness of $[t_A, t_B]$ is straightforward and omitted here.   □

Proof of Theorem 3.2.1   (i) Because of Lemma 3.2.3 it is sufficient to prove the property (3.2.13) for the radius of convergence $r(t)$ defined in (3.2.1).

Let $t_0 \in [t_A, t_B]$ be arbitrarily fixed. The KKT conditions for $P(t)$ can be written in the form of a generalized equation:

$$(E(t))\, 0 \in \begin{bmatrix} D_x f(x, t) + \sum_{i \in I} \lambda_i D_x h_i(x, t) + \sum_{j \in J} \mu_j D_x g_j(x, t) \\[2mm] - h_i(x, t), i \in I \\[2mm] - g_j(x, t), j \in J \end{bmatrix} + \partial \Psi_c(v) \qquad (3.2.14)$$

where $\partial \Psi_c$ is as in (3.2.10) with $C := \mathbb{R}^n \times \mathbb{R}^m \times \mathbb{R}^s_+$.

Analogously, the KKT conditions for $P(\bar{v}, t)$ can be written in the form

$$(E(\bar{v}, t)) \qquad\qquad \cdot 0 \in F(v, \bar{v}, t) + \partial \Psi_c(v) \qquad\qquad (3.2.15)$$

with

$$F(v, \bar{v}, t) := \begin{bmatrix} D_x \varphi(x, \bar{v}, t) + \sum_{i \in I} \lambda_i D_x h_i(\bar{x}, t) + \sum_{j \in J} \mu_j D_x g_j(\bar{x}, t) \\[2mm] - [h_i(\bar{x}, t) + D_x h_i(\bar{x}, t)(x - \bar{x})], i \in I \\[2mm] - [g_j(\bar{x}, t) + D_x g_j(\bar{x}, t)(x - \bar{x})], j \in J \end{bmatrix}. \qquad (3.2.16)$$

By [190], Theorem 4.1, the generalized equation corresponding to the KKT conditions of a nonlinear optimization problem is strongly regular at a solution (x_0, λ_0, μ_0) if the strong second-order sufficient condition holds at (x_0, λ_0, μ_0) together with the linear independence of the gradients of the active constraints. As we have assumed (V1)–(V3), it follows that $E(t_0)$ is strongly regular at $v(t_0)$.

Because of (Â1)–(Â3) the linearizations with respect to v (in the sense of (3.2.10) with $v(t_0)$ instead of x_0) of $E(t_0)$ and $E(v(t_0, t_0))$ coincide. Hence $v(t_0)$

solves $E(v(t_0, t_0))$ and by Definition 3.2.2 this generalized equation is strongly regular at $v(t_0)$, too.

Moreover, $F(v, \bar{v}, t)$ is continuous and continuously differentiable with respect to v in a neighbourhood of $(v(t_0), v(t_0), t_0)$, since F depends on (λ, μ) only linearly, and all functions contained in F and their derivatives with respect to x are continuous because of (V1) and (Â1). Thus the assumptions of Theorem 3.2.2 are fulfilled. The application of this theorem yields (we take $(\bar{v}, t)$ as the parameter p, $(v(t_0), t_0)$ instead of p_0, $U_R(v(t_0)) \times U(t_0)$ instead of N, $U_\sigma(v(t_0))$ instead of W and $\varepsilon := \alpha$): there exist an open neighbourhood $U(t_0)$, numbers $R > 0$, $\sigma > 0$ and a single-valued function $\hat{v}: U_R(v(t_0)) \times U(t_0) \to U_\sigma(v(t_0))$ such that

(a) $\hat{v}(\bar{v}, t)$ is a unique solution in $U_\sigma(v(t_0))$ of the inclusion

$$(E(\bar{v}, t)) \qquad \cdot 0 \in F(v, \bar{v}, t) + \partial \Psi_c(v), \qquad \forall \bar{v} \in U_R(v(t_0)), \quad \forall t \in U(t_0);$$

$$(b) \qquad \| \hat{v}(\bar{v}_1, t_1) - \hat{v}(\bar{v}_2, t_2) \| \leqslant 2\alpha \| F(\hat{v}(\bar{v}_2, t_2), \bar{v}_1, t_1) - F(\hat{v}(\bar{v}_2, t_2), \bar{v}_2, t_2) \|,$$
$$\forall \bar{v}_1, \bar{v}_2 \in U_R(v(t_0)), \quad \forall t_1, t_2 \in U(t_0). \qquad (3.2.17)$$

(ii) $v(.)$ is continuous (cf. Lemma 3.2.1). Consequently we can find a neighbourhood $U^1(t_0)$ so that $v(t) \in U_R(v(t_0))$ and $v(t) \in U_0(v(t_0))$, $\forall t \in U^1(t_0)$. $v(t)$ is a solution of $E(v(t), t)$.

Moreover, (a) implies that $\hat{v}(v(t), t)$ is the unique solution in $U_\sigma(v(t_0))$ of $E(v(t), t)$ for $t \in U^1(t_0) \subseteq U(t_0)$ (since $v(t) \in U_R(v(t_0))$). Summarizing these facts we see that

$$\hat{v}(v(t), t) = v(t) \qquad \forall t \in U^1(t_0). \qquad (3.2.18)$$

Then, together with (3.2.17), we obtain

$$\| \hat{v}(\bar{v}, t) - v(t) \| = \| \hat{v}(\bar{v}, t) - \hat{v}(v(t), t) \| \leqslant 2\alpha \| F(\hat{v}(v(t), t), \bar{v}, t) - F(\hat{v}(v(t), t), v(t), t) \|$$
$$\leqslant 2\alpha \| F(v(t), \bar{v}, t) - F(v(t), v(t), t) \|,$$
$$\forall \bar{v} \in U_R(v(t_0)), \quad \forall t \in U^1(t_0). \qquad (3.2.19)$$

(iii) Now we evaluate the right-hand side of (3.2.19):

$$\| F(v(t), \bar{v}, t) - F(v(t), v(t), t) \|$$

$$\leqslant \left\| D_x \varphi(x(t), \bar{v}, t) + \sum_{i \in I} \lambda_i(t) D_x h_i(\bar{x}, t) + \sum_{j \in J} \mu_j(t) D_x g_j(\bar{x}, t) \right.$$

$$\left. - \left(D_x \varphi(x(t), v(t), t) + \sum_{i \in I} \lambda_i(t) D_x h_i(x(t), t) + \sum_{j \in J} \mu_j(t) D_x g_j(x(t), t) \right) \right\|$$

$$+ \| c(\bar{x}, t) + D_x c(\bar{x}, t)[x(t) - \bar{x}] - c(x(t), t) \|, \qquad (3.2.20)$$

(where $c(x, t) = (h_1(x, t), \ldots, h_m(x, t), g_1(x, t), \ldots, g_s(x, t))$, and $D_x c(x, t)$ is the $(m + s) \times n$ matrix whose rows are the gradients with respect to x of the component functions of $c(x, t)$) because of (3.2.16) and the triangle inequality.

Further

$$\left\| D_x\varphi(x(t),\bar{v},t) + \sum_{i\in I}\lambda_i(t)D_xh_i(\bar{x},t) + \sum_{j\in J}\mu_j(t)D_xg_j(\bar{x},t) \right.$$

$$\left. -\left(D_x\varphi(x(t),v(t),t) + \sum_{i\in I}\lambda_i(t)D_xh_i(x(t),t) + \sum_{j\in J}\mu_j(t)D_xg_j(x(t),t)\right)\right\|$$

$$= \left\| D_xf(\bar{x},t) + \sum_{i\in I}\lambda_i(t)D_xh_i(\bar{x},t) + \sum_{j\in J}\mu_j(t)D_xg_j(\bar{x},t) \right.$$

$$\left. -\left(D_xf(x(t),t) + \sum_{i\in I}\lambda_i(t)D_xh_i(x(t),t) + \sum_{j\in J}\mu_j(t)D_xg_j(x(t),t)\right)\right.$$

$$\left. + D_x\varphi(x(t),\bar{v},t) - D_x\varphi(\bar{x},\bar{v},t) \right\|$$

which because of ($\hat{A}$2)

$$= \| D_xL(\bar{x},\lambda(t),\mu(t),t) - D_xL(x(t),\lambda(t),\mu(t),t) - D_x^2L(x(t),\lambda(t),\mu(t),t)[\bar{x}-x(t)]$$
$$+ D_x\varphi(x(t),\bar{v},t) - D_x\varphi(\bar{x},\bar{v},t) + D_x^2\varphi(x(t),v(t),t)[\bar{x}-x(t)] \|$$

and because of ($\hat{A}$3) and the definition of L

$$\leqslant \| D_xL(\bar{x},\lambda(t),\mu(t),t) - D_xL(x(t),\lambda(t),\mu(t),t) - D_x^2L(x(t),\lambda(t),\mu(t),t)[\bar{x}-x(t)] \|$$
$$+ \| D_x\varphi(x(t),\bar{v},t) - D_x\varphi(\bar{x},\bar{v},t) + D_x^2\varphi(x(t),\bar{v},t)[\bar{x}-x(t)] \|$$
$$+ \| D_x^2\varphi(x(t),v(t),t) - D_x^2\varphi(x(t),\bar{v},t) \| \cdot \| x-x(t) \|. \tag{3.2.21}$$

The inequalities (3.2.20) and (3.2.21) together with ($\hat{A}$1) and Lemma 3.2.2 provide: For any $\beta>0$ there exist neighbourhoods $U_{R_1}(v(t_0))$ and $U^2(t_0)$ such that

$$\| F(v(t),\bar{v},t) - F(v(t),v(t),t) \| \leqslant \beta \| \bar{x}-x(t) \|, \qquad \forall \bar{z}\in U_{R_1}(v(t_0)), \quad \forall t\in U^2(t_0). \tag{3.2.22}$$

Now we apply (3.2.22) with $\beta:=1/(4\alpha)$ and, if necessary, we restrict R_1 and $U^2(t_0)$ so that $R_1\leqslant R$, $U^2(t_0)\subseteq U^1(t_0)$. Together with (3.2.19) we obtain

$$\| \hat{v}(\bar{v},t)-v(t) \| \leqslant \tfrac{1}{2}\| \bar{x}-x(t) \| \tag{3.2.23}$$

$$\leqslant \tfrac{1}{2}\| \bar{v}-v(t) \|, \qquad \forall \bar{v}\in U_{R_1}(v(t_0)), \quad \forall t\in U^2(t_0). \tag{3.2.24}$$

(iv) We choose $R_2>0$ so that

$$R_2\leqslant \sigma/4 \qquad \text{and} \qquad R_2\leqslant R_1/2 \tag{3.2.25}$$

and $U^3(t_0)\subseteq U^2(t_0)$ such that

$$v(t)\in U_{R_2}(v(t_0)), \qquad \forall t\in U^3(t_0). \tag{3.2.26}$$

The inclusion (3.2.26) together with (3.2.24) provides

$$U_{R_2}(v(t)) \subseteq U_{R_1}(v(t_0)) \qquad \text{for } t \in U^3(t_0),$$

and thus (3.2.24) implies

$$\|\hat{v}(\bar{v}, t) - v(t)\| \leqslant \tfrac{1}{2}\|\bar{v} - v(t)\|, \qquad \forall \bar{v} \in U_{R_2}(v(t)), \quad \forall t \in U^3(t_0). \quad (3.2.27)$$

Now we prove:

$$\text{For } \bar{v} \in U_{R_2}(v(t)), t \in U^3(t_0), \hat{v}(\bar{v}, t) \text{ is the solution of}$$
$$E(\bar{v}, t) \text{ that is closest in norm to } \bar{v}. \qquad\qquad (3.2.28)$$

Let us assume that the opposite is true: v' $(v' \neq \hat{v}(\bar{v}, t))$ shall be a solution of $E(\bar{v}, t)$ with $\|v' - \bar{v}\| \leqslant \|\hat{v}(\bar{v}, t) - \bar{v}\|$. We obtain

$$\begin{aligned}
\|v' - \bar{v}\| &= \|\hat{v}(\bar{v}, t) - \bar{v}\| \\
&\leqslant \|\hat{v}(\bar{v}, t) - v(t)\| + \|v(t) - \bar{v}\| \\
&\leqslant \tfrac{3}{2}\|\bar{v} - v(t)(\| \qquad\qquad\qquad (3.2.29)
\end{aligned}$$

because of (3.2.27). Further

$$\begin{aligned}
\|v' - v(t_0)\| &\leqslant \|v' - \bar{v}\| + \|\bar{v} - v(t)\| + \|v(t) - v(t_0)\| \\
&\leqslant \tfrac{5}{2}\|\bar{v} - v(t)\| + \|v(t) - v(t_0)\|
\end{aligned}$$

because of (3.2.29),

$$\|v' - v(t_0)\| \leqslant \tfrac{5}{2}\sigma/4 + \sigma/4 < \sigma$$

(cf. (3.2.25) and (3.2.26)).

As $\hat{v}(\bar{v}, t)$ is the unique solution of $E(\bar{v}, t)$ in $U_\sigma(v(t_0))$, it follows that $v' = \hat{v}(\bar{v}, t)$, which contradicts the assumption.

(v) We recall the structure of algorithm (A) presented in Section 3.1. Let us assume that $v^0 \in U_{R_2}(v(t))$ and $t \in U^3(t_0)$. From assertion (3.2.28) we conclude: If $v^i \in U_{R_2}(v(t))$, then one step of algorithm (A) provides $v^{i+1} = \hat{v}(v^i, t)$. Because of (3.2.27) the inequality

$$\|v^{i+1} - v(t)\| \leqslant \tfrac{1}{2}\|v^i - v(t)\| \qquad\qquad (3.2.30)$$

holds. In particular, we see that v^{i+1} lies in $U_{R_2}(v(t))$ again. Thus (A) starting with v^0 provides a sequence $\{v^i\}$ satisfying (3.2.30) for $i = 0, 1, 2, \ldots$, i.e. a sequence converging to $v(t)$.

In other words, the radius of convergence of (A) solving $P(t)$ in a neighbourhood of $v(t)$ is not less than R_2:

$$r(t) \geqslant R_2 \qquad \forall t \in U^3(t_0).$$

This finishes the proof. $\qquad\qquad\qquad\qquad\qquad\qquad\qquad\qquad\qquad\qquad\Box$

REMARK 3.2.1   Applying (3.2.22) we could choose β arbitrarily small and obtain (3.2.23) and (3.2.24) with any positive factor ε instead of $\frac{1}{2}$. So, we can prove the Q-superlinear convergence rate of (A) solving $P(t)$. An immediate consequence is the R-superlinear convergence of $\{v^i\}$ and $\{x^i\}_{i=0,1,2,\dots}$. However, applying the strengthened form of (3.2.23) it is possible to prove even a Q-superlinear convergence rate for the sequence of the primal variables $\{x^i\}$. For the notions of R-superlinear and Q-superlinear rates of convergence we refer the reader to any well known text book on numerical analysis, e.g. [169] or [209].

Now we explain the pathfollowing process outlined in Section 3.1 (cf. (3.1.4)). In order to avoid double indices we prefer the following formulation.

Given a suitable partition of $[t_A, t_B]$

$$t_A = t_0 \leqslant t_1 \leqslant \cdots \leqslant t_N = t_B \tag{3.2.31}$$

and the point $v_A = v(t_A)$, compute v_k, $k = 1, \dots, N$, as a KKT point of

$$P(v_{k-1}, t_k) \tag{3.2.32}$$

with $P(.,.)$ from (3.1.3). If there is more than one such point, choose v_k to be closest in norm to v_{k-1}. This concept implies only one step of (A) per level t_k. However, condition (3.2.31) does not exclude equal levels. Consequently, this approach is sufficiently general to cover the situation described in Section 3.1.

In practice the sequence $\{t_k\}_{k=0,\dots,N}$ is not determined previous to the computation of $\{v_k\}$, but simultaneously.

Summarizing we can state the following theorem:

THEOREM 3.2.3   (see [140]) Assume (E1), (E2), (V1), (V2) and (V3) for $P(t)$, $t \in [t_A, t_B]$, and $(\hat{A}1)$, $(\hat{A}2)$ and $(\hat{A}3)$ for the locally convergent algorithm (A).

Then there exist real numbers $\Delta > 0$ and $\varepsilon > 0$ such that the algorithm described by (3.2.31) and (3.2.32), starting at the point $v_A = v(t_A)$, generates points v'_k, $k = 0, 1, \dots, N$ with $\|v'_k - v(t_k)\| < \varepsilon$, $t_k - t_{k-1} \leqslant \Delta$, $k = 1, \dots, N$ in a finite number of iteration steps.

Proof   We consider the sequence $\{v'_k\}$, $k = 0, 1, \dots, N$, determined as follows:

$$\|v'_0 - v(t_A)\| < \varepsilon, \tag{3.2.33}$$

$$\|v'_k - v_k\| < \varepsilon', \qquad k = 1, \dots, N, \tag{3.2.34}$$

where v_k is a KKT point of $P(v'_{k-1}, t_k)$ that is closest in norm to v'_{k-1} and $\varepsilon' > 0$.

From the proof of Theorem 3.2.1, especially (3.2.30) together with Lemma 3.2.3, it follows that there is a constant $r > 0$ such that

$$\|v_{k+1} - v(t)\| \leqslant \tfrac{1}{2}\|v_k - v(t)\|, \qquad \forall v_k \in U_r(v(t)), \quad \forall t \in [t_A, t_B], \tag{3.2.35}$$

where v_{k+1} is a KKT point of $P(v_k, t)$ that is closest in norm to v_k. We choose $\varepsilon' > 0$ and $\eta > 0$ such that

$$2\varepsilon' + \eta \leqslant r/2. \tag{3.2.36}$$

Because of Lemma 3.2.1 the function $v(\cdot)$ is continuous on the compact set $[t_A, t_B]$; hence there is a $\Delta > 0$ such that

$$\|v(\bar{t}) - v(t)\| < \eta \qquad \text{for all } \bar{t}, t \in [t_A, t_B] \text{ with } \|\bar{t} - t\| \leqslant \Delta. \tag{3.2.37}$$

The choice of v'_0 as in (3.2.33) is possible according to (E2). Then $\|v'_0 - v(t_A)\| < r/2$ because of (3.2.36). Given t_k and v'_k with $\|v'_k - v(t_k)\| < r/2$, for some $k \in \{0, \ldots, N-1\}$ we obtain (regarding $0 \leqslant t_{k+1} - t_k \leqslant \Delta$):

$$\|v'_k - v(t_{k+1})\| \leqslant \|v'_k - v(t_k)\| + \|v(t_k) - v(t_{k+1})\| < r/2 + \eta < r$$

because of (3.2.37) and (3.2.36). Hence (3.2.35) is applicable with v'_k instead of v_k and t_{k+1} instead of t:

$$\|v_{k+1} - v(t_{k+1})\| \leqslant \tfrac{1}{2}\|v'_k - v(t_{k+1})\| < \tfrac{1}{2}(r/2 + \eta).$$

Further we obtain:

$$\begin{aligned}
\|v'_{k+1} - v(t_{k+1})\| &\leqslant \|v'_{k+1} - v_{k+1}\| + \|v_{k+1} - v(t_{k+1})\| \\
&\leqslant \varepsilon' + \tfrac{1}{2}(r/2 + \eta) \\
&\leqslant r/2. \tag{3.2.38}
\end{aligned}$$

because of (3.2.36).

From the construction of $\{t_k\}$ and $\{v'_k\}$ it is clear that for some N, $t_N = t_B$ and $\|v'_N - v(t_B)\| < r/2$, which implies that v'_N lies in the domain of convergence of (A) for $P(t_B)$. Therefore, if we take $\varepsilon = r/2$ in (3.2.38), then the theorem is proved.

$\square$

REMARK 3.2.2 Suppose, in the algorithm (A), we consider a ciritical point instead of a KKT point and modify the assumptions in the following way:

(E1') There exists a continuous function $x: [t_A, t_B] \to \mathbb{R}^n$ such that $x(t)$ is a critical point for $P(t)$. (Note that for a critical point the LICQ is fulfilled. Therefore, the assumption (V2) is satisfied.)

(V3') $D_x^2 L(z(t)) \mid T_{x(t)}^{\neq} M(t)$ is non-singular for all $t \in [t_A, t_B]$ with $z(t)$ and $L(z)$ as in (V3), but

$$J^{\neq}(z(t)) = \{ j \in J_0(z(t)) \mid \mu_j(t) \neq 0 \}$$

$$T_{x(t)}^{\neq} M(t) = \bigcap_{i \in I} \operatorname{Ker} D_x h_i(z(t)) \cap \bigcap_{j \in J^{\neq}(z(t))} \operatorname{Ker} D_x g_j(z(t)).$$

Then, the radius of convergence can be estimated on $[t_A, t_B]$ in the same way as in Theorem 3.2.1. As a consequence, a modified procedure to follow a curve of critical points with the property (V3′) is possible, too. Furthermore, note that the assumptions (E1′) and (V3′) are fulfilled if $(f, H, G) \in \mathscr{F}^{**}$ and $z(t) \in \Sigma^1_{loc} \cap \Sigma^2_{loc}$ $(\Sigma^i_{loc} := \Sigma^i_{gc} \cap \Sigma_{loc})$.

3.3   AN ACTIVE INDEX SET STRATEGY

In this section we follow Guddat *et al.* [93] and Gfrerer *et al.* [70]. We assume (E1), (E2), (V1), (V2) and (V3). First, we introduce the notion of a local stability set. Let $\tilde{J} \subset \{1, \ldots, s\}$. We consider the parametric optimization problem

$$P^{\tilde{J}}(t): \qquad \min\{f(x,t) \mid h_i(x,t) = 0, i \in I, g_j(x,t) = 0, j \in \tilde{J}\}, \qquad t \in [t_A, t_B],$$

with only equations as constraints. The following definition connects the solution $x(t)$ of $P(t)$ with $P^{\tilde{J}}(t)$.

DEFINITION 3.3.1   Let $t_0 \in [t_A, t_B]$, $J_0 := J_0(x(t_0), t_0)$. The maximally connected subset of $\{t \in [t_A, t_B] \mid x(t)$ is a local minimizer of $P^{J_0}(t)\}$ to which t_0 belongs is called the *local stability set* with respect to $x(t)$ and t_0, denoted by $S(t_0)$.

This definition does not necessarily mean that no other index $j \in J \backslash J_0$ is allowed to become active while working with the fixed index set J_0. Any index may join the index set J_0—which of course remains active—as long as $x(t)$ remains a local minimizer of $P^{J_0}(t)$.

Let us give a first characterization of $S(t_0)$.

THEOREM 3.3.1   Assume (E1), (V1), (V2) and (V3) to be satisfied. Then the local stability set $S(t_0)$ is the maximally connected component of

$$\{t \in [t_A, t_B] \mid J^+(x(t), t) \subset J_0(x(t_0), t_0) \subset J_0(x(t), t)\},$$

to which t_0 belongs.

Proof   We show that $x(t)$ is a local minimizer for $P^{J_0}(t) \Leftrightarrow J^+(x(t), t) \subset J_0(x(t_0), t_0) \subset J_0(x(t), t)$.

(1) $\Rightarrow$: If $x(t)$ is a local minimizer for $P^{J_0}(t)$, then the KKT conditions are satisfied for $P^{J_0}(t)$.

Since $g_j(x(t), t) = 0, j \in J_0$ holds, it follows that $J_0 \subset J_0(x(t), t)$, and by (V2) we have

$$D_x f(x(t), t) + \sum_{i \in I} \lambda_i^{J_0}(t) D_x h_i(x(t), t) + \sum_{j \in J_0} \mu_i^{J_0}(t) D_x g_j(x(t), t) = 0. \qquad (3.3.1)$$

Analogously, by (E1), (V1) and (V2), $x(t)$ has to satisfy the KKT conditions for $P(t)$:

$$D_x f(x(t), t) + \sum_{i \in I} \lambda_i(t) D_x h_i(x(t), t) + \sum_{j \in J^+(x(t),t)} \mu_j(t) D_x g_j(x(t), t) = 0.$$

Therefore, by (3.3.1), it follows that

$$\sum_{i \in I} [\lambda_i(t) - \lambda_i^{J_0}(t)] D_x h_i(x(t), t) + \sum_{j \in J^+(x(t),t)} \mu_j(t) D_x g_j(x(t), t) - \sum_{j \in J_0} \mu_j^{J_0}(t) D_x g_j(x(t), t) = 0.$$

From here, we have

$$\sum_{i \in I} [\lambda_i(t) - \lambda_i^{J_0}(t)] D_x h_i(x(t), t) + \sum_{j \in J^+(x(t),t) \cap J_0} [\mu_j(t) - \mu_j^{J_0}(t)] D_x g_j(x(t), t)$$

$$+ \sum_{j \in J^+(x(t),t) \setminus J_0} \mu_j(t) D_x g_j(x(t), t) - \sum_{j \in J_0 \setminus J^+(x(t),t)} \mu_j^{J_0}(t) D_x g_j(x(t), t) = 0.$$

We note that

$$[J^+(x(t), t) \cap J_0] \cup [J^+(x(t), t) \setminus J_0] \cup [J_0 \setminus J^+(x(t), t)] - J^+(x(t), t) \cup J_0 \subset J_0(x(t), t).$$

By (V2) all gradients are linearly independent. Therefore, in particular, $u_j(t) = 0$ for $j \in J^+(x(t), t) \setminus J_0$, which implies that $J^+(x(t), t) \setminus J_0 = \varnothing$. Then $J^+(x(t), t) \subset J_0$.

$(2) \Leftarrow:$ $J_0 \subset J_0(x(t), t)$ implies that $g_j(x(t), t) = 0, j \in J_0$, and $\{D_x h_i(x(t), t), D_x g_j(x(t), t), i \in I, j \in J_0\}$ is linearly independent.

$J^+(x(t), t) \subset J_0$ and $\mu_j(t) = 0$, $j \in J_0 \setminus J^+(x(t), t)$ imply together with (E1), (V1) and (V2) that

$$D_x f(x(t), t) + \sum_{i \in I} \lambda_i(t) D_x h_i(x(t), t) + \sum_{j \in J^+(x(t),t)} \mu_j(t) D_x g_j(x(t), t)$$

$$+ \sum_{j \in J_0 \setminus J^+(x(t),t)} \mu_j(t) D_x g_j(x(t), t) = 0.$$

This means that the KKT conditions for $P^{J_0}(t)$ are satisfied. Because of $J^+(x(t), t) \subset J_0$ we have

$$\bigcap_{j \in J_0} \operatorname{Ker} D_x g_j(x(t), t) \subset \bigcap_{j \in J^+(x(t),t)} \operatorname{Ker} D_x g_j(x(t), t).$$

Hence the second-order sufficient conditions (cf. (V3)) hold for $P^{J_0}(t)$. $\qquad\square$

LEMMA 3.3.1   Assume (E1), (V1), (V2) and (V3). Then the local stability set $S(t_0)$ is closed for all $t_0 \in [t_A, t_B]$.

Proof Let $t \in S)(t_0)$ be arbitrary. From the definition of $S(t_0)$ it follows that

$$h_i(x(t), t) = 0, \quad i \in I, \qquad g_j(x(t), t) = 0, \quad j \in J_0. \tag{3.3.2}$$

Hence $J_0 \subset J(x(t), t)$. Using (V2) this implies that

$$D_x h_i(x(t), t), \quad i \in I, \qquad D_x g_j(x(t), t), \quad j \in J_0$$
$$\text{are linearly independent.} \tag{3.3.3}$$

Therefore, there exist unique multipliers $\lambda_i^{J_0}(t)$, $\mu_j^{J_0}(t)$, $i \in I$, $j \in J_0$, solving the linear equation

$$D_x f(x(t), t) + \sum_{i \in I} \lambda_i \cdot D_x h_i(x(t), t) + \sum_{j \in J_0} \mu_j \cdot D_x g_j(x(t), t) = 0. \tag{3.3.4}$$

Equations (3.3.2) and (3.3.4) are the KKT conditions for $P^{J_0}(t)$. The continuity of $x(t)$ and (V2) imply that (3.3.2) and (3.3.3) are also fulfilled for $t \in \overline{S(t_0)}$, and therefore (3.3.4) has a unique solution $\lambda_i^{J_0}(t)$, $\mu_i^{J_0}(t)$, $i \in I$, $j \in J_0$, for each $t \in \overline{S(t_0)}$, and $\lambda_i^{J_0}, \mu_j^{J_0}, i \in I, j \in J_0$, are continuous on $\overline{S(t_0)}$. Hence, for each $t \in \overline{S(t_0)}$, $x(t)$, $\lambda_i^{J_0}(t)$, $\mu_j^{J_0}(t)$, $i \in I$, $j \in J_0$, fulfil the KKT conditions for $P^{J_0}(t)$.

Trivially, $x(t), \lambda_i^{J_0}(t), \mu_j^{J_0}(t), i \in I, j \in J$ (where $\mu_j^{J_0} := 0$ for $j \notin J_0$) satisfy the equation

$$D_x f(x(t), t) + \sum_{i \in I} \lambda_i \cdot D_x h_i(x(t), t) + \sum_{j \in J} \mu_j \cdot D_x g_j(x(t), t) = 0.$$

Since $x(t), \lambda_i(t), \mu_j(t), i \in I, j \in J$, also satisfy this equation, (V2) implies $\mu_j(t) = \mu_j^{J_0}(t)$, $j \in J_0$, $\mu_j(t) = 0$, $j \notin J_0$, $\lambda_i(t) = \lambda_i^{J_0}(t)$, $i \in I$.

Then it follows from (V3) that $x(t)$, $\mu_j^{J_0}(t)$, $j \in J_0$, $\lambda_i^{J_0}(t)$, $i \in I$, also fulfil the second-order sufficient conditions (V3) for $P^{J_0}(t)$. Consequently, $x(t)$ is a local minimizer of $P^{J_0}(t)$, i.e. $t \in S(t_0)$ for each $t \in \overline{S(t_0)}$, which completes the proof.

$\square$

In the case $(f, H, G) \in \mathscr{F}^{**}$ and $(x(t), t) \in \Sigma_{\text{loc}}^1 \cup \Sigma_{\text{loc}}^2$ for all $t \in [t_A, t_B]$ (cf. Section 3.1) we know from Section 2.5 that $J_0(x(t), t) = J_0$ for all $t \in S(t_0)$.

Furthermore, if $S(t_0) = [\underline{t}, \overline{t}]$ is a finite interval with $t_A < \underline{t} < \overline{t} < t_B$, the active index set will be changed exactly by one index at $\underline{t}$ and $\overline{t}$.

DEFINITION 3.3.2 The boundary points of the local stability set are called *transition points*.

For theoretical as well as computational purposes we need the existence of a function of local minimizers on an open set containing $S(t_0)$.

LEMMA 3.3.2   Assume (E1), (V1), (V2) and (V3). Then there exist a maximally connected relatively open subset $D(t_0)$ of $[t_A, t_B]$ with $S(t_0) \subset D(t_0)$ and C^1 functions

$$x^{J_0} : D(t_0) \to \mathbb{R}^n, \qquad \lambda^{J_0} : D(t_0) \to \mathbb{R}^m, \qquad \mu^{J_0} : D(t_0) \to \mathbb{R}^{|J_0|}$$

with the following properties:

(i) $x^{J_0}(t) = x(t), \lambda^{J_0}(t) = \lambda(t), \mu_j^{J_0}(t) = \mu_j(t), j \in J_0, t \in S(t_0)$.

(ii) $D_x h_i(x^{J_0}(t), t), i \in I, D_x g_j(x^{J_0}(t), t), j \in J_0$ are linearly independent for each $t \in D(t_0)$.

(iii) $(x^{J_0}(t), \lambda^{J_0}(t), \mu^{J_0}(t))$ satisfies the KKT conditions for $P^{J_0}(t)$.

(iv) $(x^{J_0}(t), \lambda^{J_0}(t), \mu^{J_0}(t))$ is an isolated local minimizer for $P^{J_0}(t), t \in D(t_0)$.

Proof $x(t)$ is a solution of $P^{J_0}(t)$ with the corresponding multipliers $\lambda_i(t) = \lambda_i^{J_0}(t)$, $i \in I$, $\mu_j(t) = \mu_j^{J_0}(t), j \in J_0$, for each $t \in S(t_0)$; see the proof of Lemma 3.3.1. The extension of these functions in the sense stated above follows directly from Robinson [189], Theorem 2.1, applied to $P^{J_0}(t)$. $\qquad\square$

Now we give a characterization of local stability sets.

THEOREM 3.3.2   Assume (E1), (V1), (V2) and (V3). Then $S(t_0)$ is the maximally connected subset of

$$\{t \in D(t_0) : h_i(x^{J_0}(t), t) = 0, i \in I, g_j(x^{J_0}(t), t) \leq 0, j \notin J_0, \mu_j^{J_0}(t) \geq 0, j \in J_0\},$$

to which t_0 belong.

Proof   Let A be the maximally connected subset of

$$\{t \in D(t_0) : h_i(x^{J_0}(t), t) = 0, i \in I, g_j(x^{J_0}(t), t) \leq 0, j \notin J_0, \mu_j^{J_0}(t) \geq 0, j \in J_0\},$$

to which t_0 belongs. It is easy to see that the inclusion $S(t_0) \subset A$ holds because, for arbitrary $t \in S(t_0)$, Lemma 3.3.2(i) implies that $x(t) = x^{J_0}(t), \mu_j(t) = \mu_j^{J_0}(t), j \in J_0$, $\lambda_i(t) = \lambda_i^{J_0}(t), i \in I$. Hence, we have $h_i(x^{J_0}(t), t) = 0, i \in I, g_j(x^{J_0}(t), t) \leq 0, j \notin J_0$, and $\mu_j^{J_0}(t) \geq 0, j \in J_0$.

Now we assume $S(t_0)$ to be a proper subset of A. Then there exist a boundary point $\bar{t}$ of $S(t_0)$ and a connected neighbourhood N of $\bar{t}$ in $[t_A, t_B]$ with $N' \not\subset S(t_0)$ and $N' \subset A$ for each connected neighbourhood N' of $\bar{t}$ with $N' \subset N$.

It is easy to see that $x^{J_0}(t), \lambda_i^{J_0}(t), i \in I, \mu_j^{J_0}(t), j \in J$ (where $\mu_j^{J_0} := 0$ for $j \notin J_0$) as well as $x(t), \lambda_i(t), i \in I, \mu_j(t), j \in J$, fulfil the KKT conditions for $P(t), t \in N$. However, Robinson [190, Theorems 2.1 and 4.1] implies the uniqueness of the KKT points in a connected neighbourhood N'' of $\bar{t}$ in T. In particular, $x(t) = x^{J_0}(t)$

for $t \in N' := N'' \cap N$. From the definition of $S(t_0)$ and Lemma 3.3.2(iv) it follows that $N' \subset S(t_0)$ in contradiction to the choice of N. $\qquad\square$

Theorem 3.3.2 gives a computational device for transition points. We restrict our investigation to a finite number of transition points, i.e. we assume:

(V4) The number of transition points in $[t_A, t_B]$ is finite.

From Section 2.5 we known that this condition is satisfied if $(f, H, G) \in \mathscr{F}^{**}$ and $(x(t), t) \in \Sigma^1_{\text{loc}} \cup \Sigma^2_{\text{loc}}$.

A consequence of the assumption (V4) is included in the following lemma.

LEMMA 3.3.3 Assume (E1), (V1), (V2), (V3) and (V4). Then (x, λ, μ): $[t_A, t_B] \to \mathbb{R}^n \times \mathbb{R}^m \times \mathbb{R}^s$ is a PC^1 function.

Proof By the implicit function theorem, it follows from Lemma 3.3.2 and (V1) that

$$x^{J_0} : D(t_0) \to \mathbb{R}^n, \qquad \lambda^{J_0} : D(t_0) \to \mathbb{R}^m, \qquad \mu^{J_0} : D(t_0) \to \mathbb{R}^{|J_0|},$$

$$\text{are continuously differentiable on } D(t_0), t_0 \in [t_A, t_B]. \tag{3.3.5}$$

If $t \in S(t_0)$, then $x(t) = x^{J_0}(t)$, $\lambda_i(t) = \lambda_i^{J_0}(t)$, $i \in I$, $\mu_j(t) = \mu_j^{J_0}(t)$, for $j \in J_0$, and $\mu_j(t) = 0$ for $j \notin J_0$. Since $[t_A, t_B]$ can be covered by a finite number of local stability sets, λ and μ are continuous. By (E1), x is also continuous, and the theorem follows from (3.3.5). $\qquad\square$

From this investigation it follows that we can use any pathfollowing method for a parameter-dependent equation system to trace the curve $\{(x^{J_0}(t), \lambda^{J_0}(t), \mu^{J_0}(t) | t \in K\}$ where K is a compact subset of $D(t_0)$ with the property $S(t_0) \subsetneqq K \subsetneqq D(t_0)$. By using the locally convergent algorithm (A) we get that the radius $r^{J_0}(t)$ of convergence for $P^{J_0}(t)$ is bounded below by a positive number r_0, i.e. we have

$$r^{J_0}(t) \geqslant r_0 \qquad \text{for all } t \in K.$$

We note that e.g. Wilson's method as well as Newton's method provide exactly the same iteration sequence. In order to give more details for the pathfollowing procedure on K, we consider the KKT system for $P^{J_0}(t)$, $t \in K$, which is written in the following compact notation:

$$F(v, t) = 0, \qquad t \in K,$$

where $v := (x^{J_0}, \lambda^{J_0}, \mu^{J_0})$.

Then we may use one of the following standard techniques:

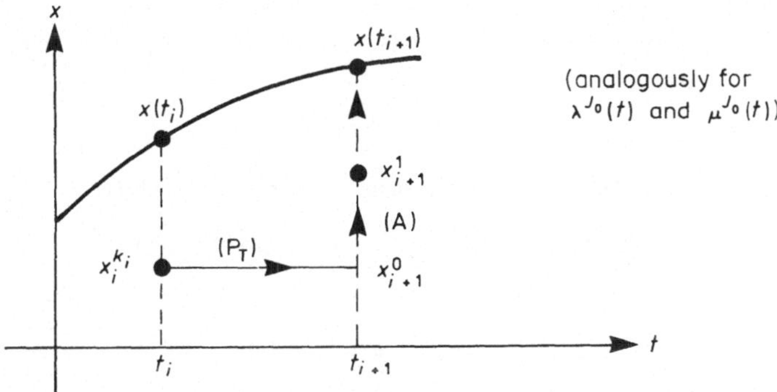

Figure 3.1

Trivial predictor

The pathfollowing method can be considered a predictor–corrector scheme of the following structure (v^k_{i+1} is the kth iterate for $v(t_{i+1})$):

(P_T) Trivial predictor: $\qquad v^0_{i+1} := v^{ki}_i$

v^{ki}_i: the last iteration for $v(t_i)$

(C) corrector: $\qquad v^{k+1}_{i+1} = A(v^k_{i+1}), \quad k = 0, 1, \ldots.$

(A) can be defined explicitly (Wilson) or implicitly (Robinson). See Figure 3.1.

Euler predictor

(P_E) Euler predictor: $v^0_{i+1} = v^{ki}_i + (t_{i+1} - t_i)\dot{v}^{ki-1}_i$. Here $\dot{v}^{ki-1}_i$ is an approximation of $\dot{v}(t_i)$ with the information based on v^{ki}_i. See Figure 3.2.

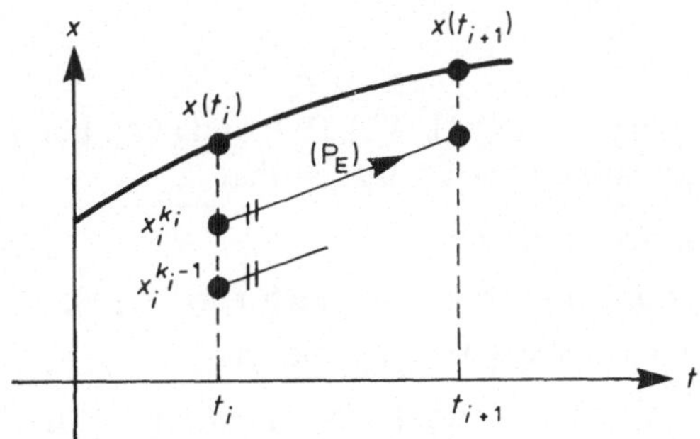

Figure 3.2

REMARK 3.3.1   In order to apply Newton's method to the nonlinear system $F(v(t), t) = 0$ we have to solve a linear system with $D_v F(v(t), t)$ as matrix. The same matrix is used to compute $\dot{v}(t)$:

$$F(v(t), t) = 0 \quad \Rightarrow \quad D_v F(v(t), t)\dot{v}(t) = -D_t F(v(t), t).$$

Therefore, using e.g. Newton's method as a corrector we have

$$D_v F(v_i^{k_i - 1}, t_i)(v_i^{k_i} - v_i^{k_i - 1}) = -F(v_i^{k_i - 1}, t_i)$$

and

$$D_v F(v_i^{k_i - 1}, t_i)\dot{v}_i^{k_i - 1} = -D_t F(v_i^{k_i - 1}, t_i).$$

Thus Euler's predictor starting at $v_i^{k_i}$ and with the tangent taken at $v_i^{k_i - 1}$ instead of $v_i^{k_i}$ does not demand essentially more computational effort than the trivial predictor at $v_i^{k_i - 1}$.

Now let us ask the following questions:

(i) How can we compute a transition point?

(ii) How can we determine the new index set of active constraints?

The answers are given in the following considerations.

Theorem 3.3.2 implicitly defines a boundary point $\bar{t}$ of $S(t_0)$: $S(t_0) = \{t \in [t_A, t_B] \mid h_i(x^{J^0}(t), t) = 0, \quad i \in I, \quad g_j(x^{J^0}(t), t) \leqslant 0, \quad j \in J, \quad \mu_j^{J^0}(t) \geqslant 0, \quad j \in J_0 \text{ with } t_0 \in S(t_0)\}$.

Therefore, in the pathfollowing process the problem of obtaining a transition point is equivalent to finding a zero of a one-dimensional functional.

Thus, to get $\bar{t}$ we may use bisection, Newton's procedure, *regula falsi* or some suitable combination of these techniques. In all cases the main computational effort has to be spent on obtaining v^{J_0}, $\dot{v}^{J_0}$ from linear systems with $D_x F$ as the matrix in both cases.

After a transition point $\bar{t}$ has been computed we have to determine the new active index set $J(\bar{t} + \varepsilon)$ for $\varepsilon > 0$ sufficiently small and $J(\bar{t})$ known.

First, we define $J^0(t) := \{ j \in J(t) \mid \mu_j(t) = 0 \}$.

THEOREM 3.3.3   Assume (E1), (V1), (V2), (V3) and (V4). Let $S(t_0) = [\underline{t}, \bar{t}], \bar{t} < t_B$. Then there exists a real number $\varepsilon > 0$ such that

(i) $J(t)$ is constant for $t \in (\bar{t}, \bar{t} + \varepsilon)$,

(ii) $J^+(\bar{t}) \subset J(t_1) \subset J(\bar{t})$ and $\bar{t} \in S(t_1)$ for each $t_1 \in (\bar{t}, \bar{t} + \varepsilon)$,

(iii) in the special case of $J^0(\bar{t}) = \{j_0\}$ one has

$$J(t) = \begin{cases} J(t_0) \cup \{j_0\} & \text{for } J(t_0) = J^+(\bar{t}), \\ J(t_0) \setminus \{j_0\} & \text{for } J(t_0) = J(\bar{t}). \end{cases}$$

Proof (i) follows obviously from (V4). (ii) follows from (i) and Theorem 3.3.1. (iii) follows from (ii). $\qquad\square$

Note that case (iii) in Theorem 3.3.3 appears under the following condition: $(f, H, G) \in \mathscr{F}^{**}$ and $(x(t), t) \in \Sigma_{\text{loc}}^1 \cup \Sigma_{\text{loc}}^2$ for all $t \in [t_A, t_B]$. In this case we do not need knowledge of the transition point. We only need the index j for which $g_j(x^{J^0}(t), t) > 0, j \in J \setminus J_0$ or $\mu_j^{J^0}(t) < 0, j \in J_0$, holds for $t > \bar{t}$ and close enough to $\bar{t}$.

With the aid of the right-sided derivative $\dot{v}(t)|_{t = \bar{t}_+}$ it is possible to sharpen the inclusion (ii) of Theorem 3.3.3 for the new active index set even in the case not covered by (iii).

THEOREM 3.3.4 Assume (E1), (V1), (V2), (V3) and (V4). Let $S(t_0) = [\underline{t}, \bar{t}], \bar{t} < t_B$. Then

(i) $\dot{x}(\bar{t}_+)$ is the unique solution of the quadratic optimization problem

$$P(t_+): \quad \min \{\tfrac{1}{2} x^{\mathsf{T}} D_x^2 L(x(\bar{t}), \lambda(\bar{t}), \mu(\bar{t}), \bar{t}) x + D_{xt}^2 L(x(\bar{t}), \lambda(\bar{t}), \mu(\bar{t}), \bar{t}) x \,|\, D_x h_i(x(\bar{t}), \bar{t}) x$$
$$+ D_t h_i(x(\bar{t}), \bar{t}) = 0, i \in I, D_x g_j(x(\bar{t}), \bar{t}) x + D_t g_j(x(\bar{t}), \bar{t}) = 0, j \in J^+(\bar{t}),$$
$$D_x g_j(x(\bar{t}), \bar{t}) x + D_t g_j(x(\bar{t}), \bar{t}) \leqslant 0, \; j \in J^0(\bar{t})\}.$$

(ii) $(\dot{\lambda}(\bar{t}_+), \dot{\mu}(\bar{t}_+))$ is the unique Lagrange multiplier related to $\dot{x}(\bar{t}_+)$ for $P(\bar{t}_+)$ with zero components for $j \notin J(\bar{t})$.

Proof (1) Uniqueness of $P(\bar{t}_+)$: Let x_s be any fixed solution of the inhomogeneous linear system

$$D_x h_i(x(\bar{t}), \bar{t}) x + D_t h_i(x(\bar{t}), \bar{t}) = 0, \quad i \in I,$$
$$D_x g_j(x(\bar{t}), \bar{t}) x + D_t g_j(x(\bar{t}), \bar{t}) = 0 \quad j \in J^+(\bar{t}), \tag{3.3.6}$$

and x_H the general solution of the homogeneous system. Then the general solution of (3.3.6) is

$$x = x_s + x_H.$$

We consider now the quadratic linearly constrained problem

$$P(t_+, x_s): \quad \min \{\tfrac{1}{2} x^{\mathsf{T}} D_x^2 L(x(\bar{t}), \lambda(\bar{t}), \mu(\bar{t}), \bar{t}) x + [D_{xt} L(x(\bar{t}), \lambda(\bar{t}), \mu(\bar{t}), \bar{t})$$
$$+ x_s^{\mathsf{T}} D_x^2 L(x(\bar{t}), \lambda(\bar{t}), \mu(\bar{t}), \bar{t})] x \,|\, D_x h_i(x(\bar{t}), \bar{t}) x = 0, i \in I,$$
$$D_x g_j(x(\bar{t}), \bar{t}) x = 0, j \in J^+(t), D_x g_j(x(\bar{t}), \bar{t}) x + D_t g_j(x(\bar{t}), \bar{t})$$
$$+ D_x g_j(x(\bar{t}), \bar{t}) x_s \leqslant 0, j \in J^0(\bar{t})\}.$$

The feasible set K for $P(\bar{t}_+, x_s)$ is convex and because of (V2) non-void. By (V3) the objective is strictly convex on K. Since we have a quadratic objective, there exists exactly one solution of $P(\bar{t}_+, x_s)$: $\tilde{x}$. Assume that $(\hat{x}, \hat{\lambda}, \hat{\mu})$ fulfils the KKT conditions for $P(\bar{t}_+, x_s)$. Then the convexity of the problem yields $\tilde{x} = \hat{x}$.

Further, the following equivalences are obvious: (x, λ, μ) is a KKT point of $P(\bar{t}_+) \Leftrightarrow (x - x_s, \lambda, \mu)$ is a KKT point of $P(\bar{t}_+, x_s)$ and x is a solution of $P(t_+) \Leftrightarrow (x - x_s)$ is a solution of $P(\bar{t}_+, x_s)$.

(2) Because of Theorem 3.3.3 there exists a $t_1 > \bar{t}$ such that $\bar{t} \in S(t_1)$ and $J(t) = \mathrm{const.}$ on $(\bar{t}, t_1]$.

We apply the implicit function theorem to the following system (the KKT conditions for $P^{J(t_1)}(\bar{t}_+)$)—see e.g. Fiacco [48]:

$$D_x f(x(\bar{t}), \bar{t}) + \sum_{i \in I} \lambda_i(\bar{t}) D_x h_i(x(\bar{t}), \bar{t}) + \sum_{j \in J(t_1)} \mu_j(\bar{t}) D_x g_j(x(\bar{t}), \bar{t}) = 0,$$

$$h_i(x(\bar{t}), \bar{t}) = 0, \qquad i \in I,$$

$$g_j(x(\bar{t}), \bar{t}) = 0, \qquad j \in J(t_1). \tag{3.3.7}$$

Because of (V1), h_i, g_j, $D_x f$, $D_x h_i$, $D_x g_j$ are continuously diffferentiable with respect to both x and t. The Jacobian of (3.3.7),

$$T := \left(\begin{array}{c|c|c} D_x^2 L(x(\bar{t}), \lambda(\bar{t}), \mu(\bar{t}), \bar{t}) & D_x^{\mathrm{T}} h_i(x(\bar{t}), \bar{t}), i \in I & D_x^{\mathrm{T}} g_j(x(\bar{t}), \bar{t}), j \in J(t_1) \\ \hline D_x h_i(x(\bar{t}), \bar{t}), i \in I & 0 & 0 \\ \hline D_x g_j(x(\bar{t}), \bar{t}), j \in J(t_1) & 0 & 0 \end{array} \right)$$

is regular on account of (V2) and (V3).

Then we have

$$\begin{pmatrix} \dot{x}^{J(t_1)}(\bar{t}_+) \\ \dot{\lambda}^{J(t_1)}(\bar{t}_+) \\ \mu^{J(t_1)}(\bar{t}_+) \end{pmatrix} = -T^{-1} \begin{pmatrix} D_{xt}^2 L(x(\bar{t}), \lambda(\bar{t}), \mu(\bar{t}), \bar{t}) \\ D_t h_i(x(\bar{t}), \bar{t}), i \in I \\ D_t g_j(x(\bar{t}), \bar{t}), j \in J(t_1) \end{pmatrix}, \tag{3.3.8}$$

and because of Lemma 3.3.2(i) it holds that

$$\dot{x}^{J(t_1)}(\bar{t}_+) = \dot{x}(\bar{t}_+), \qquad \dot{\lambda}^{J(t_1)}(\bar{t}_+) = \dot{\lambda}(\bar{t}_+), \qquad \dot{\mu}_j^{J(t_1)}(\bar{t}_+) = \dot{\mu}_j(\bar{t}_+), \qquad j \in J(t_1).$$

Further:

(i)
$$\frac{\mathrm{d}}{\mathrm{d}t} \mu_j(t)\big|_{t = \bar{t}_+} = \dot{\mu}_j(t_+) = 0$$

and

$$\frac{\mathrm{d}}{\mathrm{d}t} g_j(x(t), t)\big|_{t = \bar{t}_+} = D_x g_j(x(\bar{t}), \bar{t}) \dot{x}(\bar{t}_+) + D_t g_j(x(\bar{t}), \bar{t}) \leqslant 0$$

for $j \in J(\bar{t}) \setminus J(t_1)$,

(ii)
$$\frac{\mathrm{d}}{\mathrm{d}t} \mu_j(t)\big|_{t = \bar{t}_+} = \dot{\mu}_j(\bar{t}_+) \geqslant 0$$

and

$$\frac{d}{dt} g_j(x(t), t)\big|_{t=\bar t_+} = D_x g_j(x(\bar t), \bar t)\dot x(\bar t_+) + D_t g_j(x(\bar t), \bar t) = 0$$

for $j \in J(t_1)\setminus J^+(\bar t)$.

Because of the inclusion $J^+(\bar t) \subset J(t_1) \subset J(\bar t)$ we may extend the system (3.3.8)—arguments are suppressed to preserve clearness—to give

$D_x^2 L$	$D_x^T h_i, i\in I$	$D_x^T g_j, j\in J^+(\bar t)$	$D_x^T g_j, j\in J(t_1)\setminus J^+(\bar t)$	$D_x^T g_j, j\in J^0(\bar t)\setminus J(t_1)$
$D_x h_i, i\in I$	0	0	0	0
$D_x g_j, j\in J^+(\bar t)$	0	0	0	0
$D_x g_j, j\in J(t_1)\setminus J^+(\bar t)$	0	0	0	0
$D_x g_j, j\in J^0(\bar t)\setminus J(t_1)$	0	0	0	0

$$\times \begin{bmatrix} \dot x \\ \dot\lambda \\ \dot\mu^{J^+(\bar t)} \\ \dot\mu^{J(t_1)\setminus J^+(\bar t)} \\ \dot\mu^{J^0(\bar t)\setminus J(t_1)} \end{bmatrix} \begin{matrix} = \\ = \\ = \\ \leqslant \\ \leqslant \end{matrix} \begin{bmatrix} -D_{xt}L \\ -D_t h_i, i\in I \\ -D_t g_j, j\in J^+(\bar t) \\ -D_t g_j, j\in J(t_1)\setminus J^+(\bar t) \\ -D_t g_j, j\in J^0(\bar t)\setminus J(t_1) \end{bmatrix} \tag{3.3.9}$$

$$\dot\mu_j^{J^0(\bar t)} \geqslant 0, \qquad j\in J^0(\bar t)$$
$$\dot\mu_j^{J^0(\bar t)}(D_x g_j \cdot \dot x + D_t g_j) = 0, \qquad j\in J^0(\bar t).$$

(3) We see that (3.3.9) are the KKT conditions for $P(\bar t_+)$. $\qquad\square$

COROLLARY 3.3.1 Assume (E1), (V1), (V2) and (V3).

(1) Then

$$J^+(t) \supset J^+(\bar t) \cup \{j\in J^0(\bar t)\,|\,\dot\mu_j(\bar t_+) > 0\},$$
$$J(t) \subset J(\bar t)\setminus\{j\,|\,D_x g_j(x(\bar t), \bar t)\dot x(\bar t_+) + D_t g_j(x(\bar t), \bar t) < 0\},$$

for all $t\in(\bar t, \bar t + \delta)$, δ sufficiently small.

(2) Special case: Assume

(V5) $\quad \dfrac{d}{dt}\mu_j(t)\big|_{t=\bar t_+} \neq 0 \qquad$ or $\qquad \dfrac{d}{dt}g_j(x(t), t)\big|_{t=\bar t_+} \neq 0 \qquad$ for each $j\in J^0(\bar t)$.

Then there exists a δ' such that for all $t\in(\bar t, \bar t + \delta')$

$$J(t) = J^+(\bar t) \cup \left\{ j\in J^0(\bar t)\,\Big|\,\frac{d}{dt}g_j(x(t), t)\Big|_{t=\bar t_+} = 0 \right\}.$$

REMARK 3.3.2 (V5) is identical with the strict complementarity condition for the problem $P(\bar{t}_+)$.

Up to now our discussion has been based mainly on the knowledge of the exact solution. As this is hardly the case in reality, we analyse the situation under perturbations.

THEOREM 3.3.5 Consider the problem

$$P(t,\tau): \quad \min\{f(x,t,\tau)\,|\,h_i(x,t,\tau)=0, i\in I, g_j(x,t,\tau)\leqslant 0, j\in J\}, \qquad \tau\in\mathbb{R}^p,$$

with $P(t,0)\equiv P(t)$.

Assume (E1), (V1), (V2) and (V3) for $P(t)$ and (V1) analogously for $P(t,\tau)$: (V1)'. Then there exist numbers $\varepsilon_0>0$, $\delta_0>0$, $\alpha_0>0$ and a neighbourhood U of $\{v(t)\,|\,t\in[t_A,t_B]\}$ $(v(t)=(x(t),\lambda(t),\mu(t))$ with $x(t)$ given by (E1) and $(\lambda(t),\mu(t))$ the corresponding unique vector of Lagrange multipliers for $P(t))$ and a unique function

$$\hat{v}=(\hat{x},\hat{\lambda},\hat{\mu})\colon[t_A,t_B]\times U_{\varepsilon_0}(0)\to U,$$

where $U_r(y)$ denotes the open ball with centre y and radius r in a normed space, with the following properties:

(i) $\hat{v}(t,0)=v(t)$, $t\in[t_A,t_B]$;

(ii) for each $t\in[t_A,t_B]$ there is a $\delta(t)\geqslant\delta_0$ such that $\hat{v}(t,\tau)$ is the unique KKT point for $P(t,\tau)$ in $U_{\delta(t)}(v(t))$ for $\tau\in U_{\varepsilon_0}(0)$;

(iii) v is Lipschitz continuous on $[t_A,t_B]\times U_{\varepsilon_0}(0)$ with Lipschitz modulus α_0;

(iv) in $\hat{v}(t,\tau)$, (V2) and (V3) hold for $P(t,\tau)$, $\tau\in U_{\varepsilon_0}(0)$;

(v) $\hat{x}(t,\tau)$ is a local minimizer for $P(t,\tau)$.

Proof (1) 'Local': Let $t'\in[t_A,t_B]$. Because of (V2) and (V3) we have that $P(t',0)\equiv P(t')$ is strongly regular (Robinson [190], Theorem 4.1). Again by Robinson [190], Theorem 2.1, and (E1) we have the following: There exist $\varepsilon_1(t')>0$, $\delta_1(t')>0$, $\alpha(t')>0$ and a function $\hat{v}_{t'}\colon U_{\varepsilon_1(t')}(t')\times U_{\varepsilon_1(t')}(0)\to U_{\delta_1(t')}(v(t'))$ where the following hold:

(i) $\hat{v}_{t'}(t,0)=v(t)$, $t\in U_{\varepsilon_1(t')}(t')$.

(ii) $\hat{v}_{t'}(t,\tau)$ is a unique KKT point in $U_{\delta_1(t')}(v(t'))$ for $P(t,\tau)$, $(t,\tau)\in U_{\varepsilon_1(t')}(t,0))$ with $\|(t,\tau)\|:=\max(|t|,\|\tau\|)$.

(iii) $\hat{v}_{t'}(t,\tau)$ is Lipschitz continuous on $U_{\varepsilon_1(t')}((t',0))$ with Lipschitz modulus $\alpha(t')$.

Further we have:

(iv) There are $\varepsilon(t')$ such that, in $\hat{v}_{t'}(t,\tau)$, (V2) and (V3) hold for $P(t,\tau)$ with $(t,\tau)\in U_{\varepsilon(t')}(t',0)$.

By (V1), resp. (V1)', (V2) and (iii) we obtain: There are $\varepsilon_2(t') > 0$, $\varepsilon_2(t') \leqslant \varepsilon_1(t')$ such that for all $(t,\tau)\in U_{\varepsilon_2(t')}((t',0))$ the following hold:

(a) $g_j(\hat{x}_{t'}(t,\tau),t,\tau) < 0,\ j\notin J(t')$.

(b) $\{D_x h_i(\hat{x}_{t'}(t,\tau),t,\tau), D_x g_j(\hat{x}_{t'}(t,\tau),t,\tau), i\in I, j\in J(t')\}$ is linearly independent.

(c) $(\hat{\mu}_{t'})_j(t,\tau) > 0,\ j\in J^+(t')$.

By (a) we get $J(t') \supset J(t,\tau) := \{j\in J\,|\,g_j(\hat{x}_{t'}(t,\tau),t,\tau)=0\}$ and hence by (b) (V2) holds for $P(t,\tau)$.

To show (V3) we use the following equivalence (see Hestenes [100], p. 428):

(V3) $\Leftrightarrow$ There exists a $c > 0$ with

$$D_x^2 L(\hat{v}_{t'}(t,\tau),t,\tau) + c[D_x^T g_j(\hat{x}_{t'}(t,\tau),t,\tau), j\in J^+(t,\tau)][D_x g_j(\hat{x}_t(t,\tau),t,\tau), j\in J^+(t,\tau)]$$

$$(3.3.10)$$

being positive definite.

By assumption, (V3) holds in $v(t',0)$ for $P(t,\tau)$. Using the equivalence we have a $c > 0$ with (3.3.10). Again by (V1), resp. (V1)', and (iii) we have: There exists an $\varepsilon(t') \leqslant \varepsilon_2(t')$ such that, for all $(t,\tau)\in U_{\varepsilon(t')}((t,0))$, we have

$$D_x^2 L(\hat{v}_{t'}(t,\tau),t,\tau) + c[D_x^T g_j(\hat{x}_{t'}(t,\tau),t,\tau), j\in J^+(t')][D_x g_j(\hat{x}_{t'}(t,\tau),t,\tau), j\in J^+(t')]$$

is positive definite.

Using (c) we get $J^+(t') \subset J^+(t,\tau)$ and hence (V3) by (3.310).

(v) $\hat{v}_{t'}(t,\tau)$ fulfils the necessary and sufficient conditions. Thus, $\hat{x}_{t'}(t,\tau)$ is a local minimizer for $P(t,\tau)$.

(2) 'Global': (2.1) The set $B := \{(v(t),t)\,|\,t\in[t_A,t_B]\}$ is compact. Therefore there exists a finite partition of $[t_A,t_B]:\{t_i\},\ i=1,\dots,k, t_i\in[t_A,t_B]$, such that

$$\bigcup_{i=1}^k U_{\delta_1(t_i)}(v(t_i)) \times U_{\varepsilon(t_i)}(t_i) \supset B. \tag{3.3.11}$$

Let $\varepsilon_0 := \min\{\varepsilon(t_i)\,|\,i=1,\dots,k\}, \alpha_0 := \max\{\alpha(t_i)\,|\,i=1,\dots,k\}$. We define the function

$$\hat{v}:[t_A,t_B] \times U_{\varepsilon_0}(0) \to U := \bigcup_{t\in[t_A,t_B]} U_{\delta_1(t)}(v(t))$$

by

$$\hat{v}(t,\tau) := \hat{v}_{t_i}(t,\tau) \qquad \text{for } t\in U_{\varepsilon(t_i)}(t_i).$$

We have to show that $\hat{v}(t,\tau)$ is well defined: we use Schauder's proof strategy. Let $t_0 \in U_{\varepsilon(t_i)}(t_i) \cap U_{\varepsilon(t_j)}(t_j)$, $i \neq j$. Assume for some $\tau_0 \in U_{\varepsilon_0}(0)$: $\hat{v}_{t_i}(t_0, \tau_0) \neq \hat{v}_{t_j}(t_0, \tau_0)$. To get a contradiction we define the following homotopies:

$$q_k : [t_A, t_B] \to U_{\delta_1(t_k)}(v(t_k)) \qquad \text{by } \beta \to \hat{v}_{t_k}(t_0, \beta\tau_0), \qquad k = i, j.$$

By (1) (iii) q_i, q_j are continuous and by (1) (i) we have

$$q_i(0) = q_j(0) = v(t_0).$$

Hence

$$D := \{\beta \in [t_A, t_B] \mid q_i(\beta) = q_j(\beta)\}$$

is non-empty and compact. $\beta_0 := \max D$ and by assumption $\beta_0 < t_B$. We have

$$q_i(\lambda_0) = q_j(\lambda_0) \in U_{\delta_1(t_i)}(v(t_i)) \cap U_{\delta_1(t_j)}(v(t_j)) =: V.$$

Then $q_i^{-1}(V) \cap q_j^{-1}(V)$ is a (relatively) open subset of $[t_A, t_B]$ containing λ_0. Hence, there exists a $\tilde{\lambda} > \lambda_0$ with $q_i(\tilde{\lambda}) \neq q_j(\tilde{\lambda})$, by definition of λ_0 and $q_i(\tilde{\lambda})$, $q_j(\tilde{\lambda}) \in V$ in contradiction to (1) (ii).

Then, by construction of v, the statements (i), (iii), (iv) and (v) follow immediately.

(2.2) To get (ii) we first consider the real-valued continuous function

$$\theta : [t_A, t_B] \to \mathbb{R}^+, \qquad \theta(t) := \min\left\{\max\left\{\frac{\|v(t) - v(t_i)\|}{\delta_1(t_i)}, \frac{|t - t_i|}{\varepsilon(t_i)}\right\}\middle| i = 1, \ldots, k\right\}.$$

Then there exists a $t_0 \in [t_A, t_B]$ with $\theta(t_0) = \max\limits_{t \in [t_A, t_B]} \theta(t)$. By (3.3.11) we have $\theta(t_0) < 1$.

Let

$$\tilde{\delta} := \min\{\delta_1(t_i) \mid i = 1, \ldots, k\},$$
$$\delta_0 := [1 - \theta(t_0)]\tilde{\delta}, \qquad \delta(t) := \max\{\delta_0, \delta_1(t)\}.$$

Then $\delta(t) \geqslant \delta_0$ is obvious. It remains to show that $\delta(t)$ fulfils (ii).

Let $t \in [t_A, t_B]$, $\tau \in U_{\varepsilon_0}(0)$. Let v^1, $v^2 \in U_{\delta_0}(v(t))$ be two KKT points for $P(t, \tau)$ with $v \neq v^2$.

By construction of θ there exists an $i \in \{1, \ldots, k\}$ with $|t - t_i| \leqslant \theta(t_0)\varepsilon(t_i) < \varepsilon(t_i)$ and $\|v(t) - v(t_i)\| \leqslant \theta(t_0)\delta_1(t_1)$. Then, for $j = 1, 2$ we have

$$\|v^j - v(t_i)\| \leqslant \|v^j - v(t)\| + \|v(t) - v(t_i)\| < [1 - \theta(t_0)]\tilde{\delta} + \theta(t_0)\delta_1(t_i) \leqslant \delta_1(t_i)$$

in contradiction to (1) (ii). $\qquad\square$

COROLLARY 3.3.2 Assume (E1), (V1), (V2) and (V3). Then there exist $\delta_0 > 0$, $\alpha_0 > 0$ such that the following hold:

(i) $v(t)$ is a unique KKT point in $U_{\delta_0}(v(t))$,

(ii) $v(t)$ is Lipschitz continuous on $[t_A, t_B]$ with modulus α_0.

Theorem 3.3.5 provides a theoretical tool to analyse multiparametric optimization. In our approach we use Theorem 3.3.5 as a basis for discussing the problem of implementability of the algorithm.

In numerical reality we only get aproximations $\tilde{t}$, $\tilde{v}$, $\tilde{\dot{v}}$ for $\bar{t}$, $v(\bar{t})$, $\dot{v}(\bar{t})$ with

$$|\bar{t} - \tilde{t}| \leqslant \varepsilon_t \wedge \|\tilde{v} - v(\bar{t})\| \leqslant \varepsilon_v \wedge \|\tilde{\dot{v}} - \dot{v}(\bar{t})\| \leqslant \varepsilon_{\dot{v}},$$

where $\varepsilon_t, \varepsilon_v, \varepsilon_{\dot{v}}$ are prescribed by the algorithm. On the basis of this information we define the approximations $\tilde{J}$, $\tilde{J}^+$, $\tilde{J}^0$ for $J(\bar{t})$, $J^+(\bar{t})$, $J^0(\bar{t})$:

$$\tilde{J} := \{ j \notin J_0 \,|\, g_j(\tilde{x}, \tilde{t}) + |D_x g_j(\tilde{x}, \tilde{t})\tilde{\dot{x}} + D_t g_j(\tilde{x}, \tilde{t})|\varepsilon_t \geqslant -\varepsilon_t - [\|D_x g_j(\tilde{x}, \tilde{t})\| + 1]\varepsilon_v \} \cup J_0,$$

$$\tilde{J}^+ := \{ j \in J_0 \,|\, \tilde{\mu}_j \geqslant |\tilde{\dot{\mu}}_j|\varepsilon_t + \varepsilon_v + \varepsilon_t \}, \qquad \tilde{J}^0 := \tilde{J} \setminus \tilde{J}^+. \tag{3.3.12}$$

To apply Theorem 3.3.5 we define the following optimization problem

$$\tilde{P}(t): \quad \min \{ f(x, t) + \tilde{c}^T x \,|\, h_i(x, t) = 0, i \in I, g_j(x, t) \leqslant \tilde{b}_j, j \in J \}, \qquad t \in [t_A, t_B],$$

where

$$\tilde{c} := - D_x f(\tilde{x}, \tilde{t}) - \sum_{i \in I} \tilde{\lambda}_i D_x h_i(\tilde{x}, t) - \sum_{j \in \tilde{J}^+} \tilde{\mu}_j D_x g_j(x, \tilde{t}),$$

$$\tilde{b}_j := \begin{cases} g_j(\tilde{x}, \tilde{t}), & i \in \tilde{J}, \\ 0, & j \notin \tilde{J}. \end{cases}$$

THEOREM 3.3.6   Assume (E1), (V1), (V2) and (V3) together with the notions given above. Then, for $(\varepsilon_t, \varepsilon_v, \varepsilon_{\dot{v}})$ sufficiently small, $\tilde{P}(t)$ has the following properties:

(i) $\tilde{P}(t)$ fulfils (E1) and (V1); $\tilde{P}(\tilde{t})$ fulfils (V2) and (V3).

(ii) (a) $(\tilde{x}, \hat{\lambda}, \hat{\mu})$ with $\hat{\lambda}_i := \tilde{\lambda}_i$, $i \in I$, $\hat{\mu}_j := \tilde{\mu}_j$, $j \in \tilde{J}^+$, $\hat{\mu}_j := 0$ else, satisfies the KKT conditions for $\tilde{P}(\tilde{t})$.

(b) $\tilde{x}$ is a local minimizer for $\tilde{P}(\tilde{t})$ and $\tilde{J}^+$, $\tilde{J}^0$ are the related index sets for $\tilde{P}(\tilde{t})$ (corresponding to the index sets J^+, J^0 for $P(t)$).

Assume further that (V4) is fulfilled for $\tilde{P}(t)$. Then

(iii) (a) $\dot{\tilde{x}}(\tilde{t}_+)$ is the unique solution for the problem

$$\tilde{P}(\tilde{t}_+): \quad \min \{ \tfrac{1}{2} x^T D_x^2 L^{\tilde{J}}(\tilde{x}, \hat{\lambda}, \hat{\mu}, \tilde{t})x + D_{xt} L^{\tilde{J}}(\tilde{x}, \hat{\lambda}, \hat{\mu}, \tilde{t})x \,|\, D_x h_i(\tilde{x}, \tilde{t})x +$$
$$D_t h_i(\tilde{x}, \tilde{t}) = 0, i \in I, D_x g_j(\tilde{x}, \tilde{t})x + D_t g_j(\tilde{x}, \tilde{t}) \leqslant 0, j \in \tilde{J}^+, D_x g_j(\tilde{x}, \tilde{t})x$$
$$+ D_t g_j(\tilde{x}, \tilde{t}) \leqslant 0, j \in \tilde{J}^0 \}.$$

(b) $(\dot{\tilde{\lambda}}(\tilde{t}_+) - \dot{\tilde{\mu}}(\tilde{t}_+))$ is the corresponding unique Lagrange multiplier vector related to $\dot{\tilde{x}}(\tilde{t}_+)$ for $\bar{P}(\tilde{t}_+)$ with zero components for $j \notin \tilde{J}$.

Proof (i) follows from Theorem 3.3.5 with $\tau = (\tilde{c}, \tilde{b})$. (ii) follows from the construction of $\tilde{P}(t)$ and by Theorem 3.3.5(ii). (iii) follows from Theorem 3.3.4.

$\square$

REMARK 3.3.3   Theorem 3.3.6 can be interpreted in the following sense. It is possible to compute the new active index set exactly from the perturbed problem $\tilde{P}(t, \tau)$ with $\tau = (\tilde{c}, \tilde{b})$ if the prescribed numbers $(\varepsilon_t, \varepsilon_v, \varepsilon_{\dot{v}})$ are sufficiently small. From Theorem 3.3.5 we derive the existence of $\varepsilon_0 > 0, \delta_0 > 0, \alpha_0 > 0$ such that

$$\| \tilde{v}(t, c, b) - v(t) \| = \alpha_0 \| (c, b) \| \qquad \text{for all } t \in [t_A, t_B]$$

where $\tilde{v}(t, c, b)$ is a KKT point for

$$P(t, c, b): \qquad \min\{ f(x, t) + c^T x \mid h_i(x, t) = 0, i \in I, g_j(x, t) \leqslant b_j, j \in J \}$$

$$t \in [t_A, t_B], \quad (c, b) \in U_{\varepsilon_0}((0, 0)).$$

In particular we have: for all sequences $\{(c^i, b^i)\} \to 0$ we obtain uniform convergence of the sequence $\{\tilde{v}(\cdot, c^i, b^i)\} \to v(\cdot)$.

Next we will show that the new active index set of the perturbed problem $\tilde{P}(\tilde{t})$ provides the exact new active index set of the problem $P(t)$ with the aid of the procedure (3.3.12)

LEMMA 3.3.4   Assume (E1), (V1), (V2) and (V3). Then for all $(\varepsilon_t, \varepsilon_v, \varepsilon_{\dot{v}})$ sufficiently small we have

$$J(\tilde{t}) = \tilde{J}, \qquad J^+(\tilde{t}) = \tilde{J}^+, \qquad J^0(\tilde{t}) = \tilde{J}^0.$$

Proof (i) Let $\varphi: D \subset \mathbb{R}^p \to \mathbb{R}^q$ be continuously differentiable, and D open. Then, for all $w \in D$ and all $\varepsilon > 0$ with $\bar{U}_\varepsilon(w) \subset D$ and $w_1, w_2 \in U_\varepsilon(w)$ we have

$$\| \varphi(w_1) - \varphi(w_2) - \varphi'(w_2)(w_1 - w_2) \| = r(w_1, w_2) \| w_1 - w_2 \|,$$

where r is continuous on $\bar{U}_\varepsilon(w) \times \bar{U}_\varepsilon(w)$ and $r(w_1, w_1) = 0$.
(ii) For

$$t, t' \in U_{\varepsilon_t}(\tilde{t}) \subset D(t_0), \qquad x' \in U_{\varepsilon_v}(x(t')), \qquad \dot{x}' \in U_{\varepsilon_{\dot{v}}}(\dot{x}(t'))$$

we have: by using (i)

$$|g_j(x(t), t) - g_j(x', t') - [D_x g_j(x', t')\dot{x}' + D_t g_j(x', t')](t - t')|$$
$$\leqslant |g_j(x(t), t) - g_j(x(t'), t') - [D_x g_j(x(t'), t')\dot{x}(t') + D_t g_j(x(t'), t')](t - t')|$$

$$+ |g_j(x',t') - g_j(x(t'),t')| + |D_x g_j(x',t')\dot{x}' + D_t g_j(x',t') - D_x g_j(x(t'),t')\dot{x}(t')$$

$$- D_t g_j(x(t'),t')| \, |t - t'|$$

$$\leqslant r_1(t,t')|t - t'| + |D_x g_j(x',t')[x(t') - x']| + r_2(x',x(t'))\|x' - x(t')\|$$

$$+ [\|x' - x(t')\| \, \|D_x^2 g_j(x',t')\| \, \|\dot{x}'\| + r_3(x',x(t'))\|x' - x(t')\| \, \|\dot{x}'\|$$

$$+ \|D_x g_j(x(t'),t')\| \, \|\dot{x}(t') - \dot{x}'\| + \|D_{xt} g_j(x',t')\| \, \|x' - x(t')\|$$

$$+ r_4(x',x(t'))\|x - x(t')\| \,]|t - t'|$$

$$\leqslant |t - t'| + [\|D_x g_j(x',t')\| + 1]\|x' - x(t')\|$$

(for $(\varepsilon_t, \varepsilon_v, \varepsilon_{\dot{v}})$ sufficiently small).

Analogously we get $|\mu_j(t) - \mu'_j - \dot{\mu}'_j(t - t')| \leqslant |\mu'_j - \mu_j(t')| + |t - t'|$ (for $(\varepsilon_t, \varepsilon_v, \varepsilon_{\dot{v}})$ sufficiently small.)

(iii) Now we show that the index sets are exact. Let $j \in J(\bar{t}) \backslash J^0$. Then (ii) implies

$$|g_j(x(\bar{t}),\bar{t}) - g_j(\tilde{x},\tilde{t}) - [D_x g_j(\tilde{x},\tilde{t})\dot{\tilde{x}} + D_t g_j(\tilde{x},\tilde{t})](\tilde{t} - \bar{t})|$$

$$\leqslant |\tilde{t} - \bar{t}| + [\|D_x g_j(\tilde{x},\tilde{t})\| + 1]\|\tilde{x} - x(\bar{t})\|.$$

We obtain

$$g_j(\tilde{x},\tilde{t}) + |D_x g_j(\tilde{x},\tilde{t})\dot{\tilde{x}} + D_t g_j(\tilde{x},\tilde{t})|\varepsilon_t \geqslant -\varepsilon_t - [\|D_x g_j(\tilde{x},\tilde{t})\| + 1]\varepsilon_v$$

Now assume $j \notin J(\bar{t}) \Rightarrow g_j(x(\bar{t}),\bar{t}) =: c_j < 0$. Then, by (VI) there exists an $\varepsilon_t > 0$ such that, for all $t \in U_{\varepsilon_t}(\bar{t})$, we have $g_j(x(t),t) < c_j/2$.

$$\rightarrow \qquad g_j(\tilde{x},\tilde{t}) + |D_x g_j(\tilde{x},\tilde{t})\tilde{x} + D_t g_j(\tilde{x},\tilde{t})|\varepsilon_t$$

$$< c_j/4 + |D_x g_j(\tilde{x},\tilde{t})\dot{\tilde{x}} + D_t g_j(\tilde{x},\tilde{t})|\varepsilon_t$$

$$< -\varepsilon_t - [\|D_x g_j(\tilde{x},\tilde{t})\| + 1]\varepsilon_v$$

for $(\varepsilon_t, \varepsilon_v, \varepsilon_{\dot{v}})$ sufficiently small; that is, the criterion (3.3.12) is violated.

The proof for $\tilde{J}^+ = J^+(\bar{t})$ runs analogously. $\qquad\square$

REMARK 3.3.4 To get the new active index set after passing $\bar{t}$ we have to solve $P(\bar{t}_+)$ and, in practice, we solve an approximate problem $\tilde{P}(\bar{t}_+)$. Lemma 3.3.4 states that we have exact active index sets at the point $\bar{t}$ for $(\varepsilon_t, \varepsilon_v, \varepsilon_{\dot{v}})$ sufficiently small. Hence $\tilde{P}(\bar{t}_+)$ is a perturbation of the exact problem $P(\bar{t}_+)$. Therefore, for $(\varepsilon_t, \varepsilon_v, \varepsilon_{\dot{v}})$ sufficiently small, the regularity assumptions (V2) and (V3) together with (V1) state that the solution of the perturbed problem depends continuously on the perturbation. Hence it is easy to see that we obtain the exact new active set for $(\varepsilon_t, \varepsilon_v, \varepsilon_{\dot{v}})$ sufficiently small if (V5) holds.

3.4 THE ALGORITHM PATH I AND NUMERICAL RESULTS

In this section we describe an algorithm that is based on the theoretical results given above. We discussed some predictor–corrector schemes for the pathfollowing process before. To dispense with special techniques we confine ourselves here to the abstract structure of the algorithm. Furthermore, we assume to have the starting values x_0, λ_0, μ_0, J_0.

Algorithm PATH I

1. Given x_0, J_0, λ_0, μ_0

$$\varepsilon_t, \varepsilon_v, \varepsilon_{\dot{v}}, \Delta t_{\min}, \Delta t_{\max} \in \mathbb{R}^+$$
$$t_0 := 0$$
$$k := 1$$

2. Determination of the step size $\Delta t_k \in [\Delta t_{\min}, \Delta t_{\max}]$,

$$t_k := t_{k-1} + \Delta t_k$$

3. Compute $v = (x_k, \lambda_k, \mu_k)$ with $\| v_k - v(t_k) \| \leqslant \varepsilon_v$

$$v(t_k) = (x(t_k), \lambda(t_k), \mu(t_k))$$

4. Check:

$$g_j(x_k, t_k) < - \| D_x g_j(x_k, t_k) \| \varepsilon_v, \, j \notin J_{k-1} \wedge (\mu_k)_j > \in_v, \, j \in J_{k-1}$$

Yes: $J_k := J_{k-1}$ goto 5
No: $\Rightarrow t_k > \bar{t}$. Compute $\tilde{t} < t_k$, $\tilde{v}$, $\dot{\tilde{v}}$ with

$$|\tilde{t} - \bar{t}| \leqslant \varepsilon_{t'} \| \tilde{v} - v^{J_{k-1}}(\tilde{t}) \| \leqslant \varepsilon_v, \| \dot{\tilde{v}} - \dot{v}^{J_{k-1}}(\tilde{t}) \| \leqslant \varepsilon_{\dot{v}}$$
$$\tilde{J} := J_{k-1} \cup \{ j \notin J_{k-1} | g_j(\tilde{x}, \tilde{t}) + | D_x g_j(\tilde{x}, \tilde{t})\dot{\tilde{x}} + D_t g_j(\tilde{x}, \tilde{t})|\varepsilon_t$$
$$\geqslant - \varepsilon_t - [\| D_x g_j(\tilde{x}, \tilde{t}) \| + 1]\varepsilon_v \}$$
$$\tilde{J}^+ := \{ j \in J_{k-1} | \tilde{\mu}_j \geqslant | \dot{\tilde{\mu}}_j | \varepsilon_t + \varepsilon_v + \varepsilon_t \}$$
$$\tilde{J}^0 := \tilde{J} \backslash \tilde{J}^+$$
$$|\tilde{J}^0| = 1 \Rightarrow J_k$$
$$|\tilde{J}^0| > 1 \Rightarrow \text{Solve } P(\tilde{t}_+) \Rightarrow J_k$$

5. $k := k + 1$
   Goto 2

THEOREM 3.4.1   Assume (E1), (E2) and (V1)–(V5). Then, for all ε, $(\varepsilon_t, \varepsilon_v, \varepsilon_{\dot{v}})$, $\Delta t_{\max}$ sufficiently small, PATH I generates a discretization $t_A = t_0 < \cdots < t_i < t_{i+1} < \cdots < t_N = t_B$ and corresponding points $(\tilde{x}^i, \tilde{\lambda}^i, \tilde{\mu}^i)$, $\| (\tilde{x}^i, \tilde{\lambda}^i, \tilde{\mu}^i) - (x(t_i), \lambda(t_i), \mu(t_i)) \| < \varepsilon$, $i = 1, \ldots, N$.

Proof   The algorithm PATH I can be split into three different strategies.

(i) Pathfollowing in $D(t_0)$: see Lemma 3.3.2.

(ii) Determination of an approximation $\tilde{t}$ of a transition point $\bar{t}$: see Theorem 3.3.2.

(iii) Determination of the new active index set
Case 1: $|J^0| = 1$; see Theorem 3.3.3.
Case 2: $|J^0| > 1$; see Theorem 3.3.4, Lemma 3.3.4, Remark 3.3.4. $\qquad\qquad\square$

REMARK 3.4.1 (i) In principle, any reasonable step size control known from the application of pathfollowing methods to one-parametric nonlinear equations may be used when passing through any stability set. Some of these techniques are independent of the special local iterative method to solve $F(v, t) = 0$, e.g. the well known doubling/bisecting strategy. Other strategies depend strongly on the local method, see e.g. [34]. One can even try to minimize the total numerical effort inside a stability interval; see [97] and [242].
$(x^{J_{k-1}}(t_k), \lambda^{J_{k-1}}(t_k), \mu^{J_{k-1}}(t_k), t_k)$ is an exact point of the continuation path, which is unknown in general. The criterion in PATH I step 3 has to be made implementable by some of the well known devices discussed for instance in Schwetlick [209], p. 126. After Lemma 3.3.3 we offered some predictor corrector techniqes to obtain v_k from v_{k-1}. An Euler predictor–Newton corrector strategy seems appropriate from several points of view. Under the assumption that f, h_i, $g_i \in C^3(\mathbb{R}^n \times \mathbb{R}, \mathbb{R})$, $i \in I$, $j \in J$, we have a (local) order $O((\Delta t)^4)$, though only one Jacobian step has to be computed or approximated. Further, Zulehner's optimal step size control [242] may be used inside each stability interval.

(ii) The following alternative strategy might also be interesting. To pass over transition points one turns to the full problem $P(t)$ instead of $P^{J_0}(t)$, and implements the algorithm (A) (cf. Section 3.2), or more exactly one of its special cases, computes the new index set of active constraints, J_1, and, having left the transition region, goes back to the new reduced problem $P^{J_1}(t)$. As a further alternative strategy we finally mention the active index set strategy of Lehmann [141].

REMARK 3.4.2 If we consider, analogously to Remark 3.2.2, a critical point instead of a stationary point in the algorithm PATH I and modify the assumptions (E1), (V3) by (E1'), (V3') (cf. Remark 3.2.2), then a modified algorithm to follow a curve of critical points with the property (V3') will be possible, too.

The following examples were computed with the algorithm PATH I. For the local iterative optimization the augmented Lagrange multiplier method (cf. Gfrerer [69]) was used.

Example 3.4.1

(See [93] for an analytical discussion.)

$$P(t): \quad \min\{x_1^2 + (4t - 2)x_2 + \tfrac{1}{2}x_2^2 \,|\, g_1 = x_1^2 + x_2^2 - 1 \leqslant 0, g_2 = x_1 - x_2 - t - \tfrac{1}{4} \leqslant 0\}$$
$$[t_A, t_B] = [0, 1].$$

The example was constructed to illustrate Corollary 1 and Theorem 4 from [93]. Here (Table 3.1) it elucidates the computation of a transition point as well as the determination of the new index set.

Example 3.4.2

(See Hackl [96], where this example was solved by a globalized optimization procedure based on embedding.)

Table 3.1

t	x	x_2			$J\{t\}$
0.0	0.0	1.0	0.5	0.0	$\{1\}$
0.1	0.0	1.0	0.3	0.0	
0.2	0.0	1.0	0.1	0.0	
0.3*	0.0	1.0	-0.1	0.0	$\mu_2 \geqslant 0$ violated
0.2500	0.0	1.0	-0.0001	0.0	$\{\varnothing\}$
0.2600	0.0	0.9599	-0.0001	0.0	
0.3600	0.0	0.5599	-0.0001	0.0	
0.4600	0.0	0.1599	-0.0001	0.0	
0.5600	0.0	-0.2401	-0.0001	0.0	
0.6600	0.0	-0.6401	-0.0001	0.0	
0.7600*	0.0	-1.0401	-0.0001	0.0	$g_1(x,t) \leqslant 0, \quad g_2(x,t) \leqslant 0$ violated
0.7500	0.0	-1.0000	-0.0001	0.0	$\{1\}$
0.7600	0.0000	-1.0000	0.0000	0.0	
0.8600	0.0000	-1.0000	0.2200	0.0	
0.9600	0.0000	-1.0	0.4200	0.0	
1.0	0.0	-1.0	0.5	0.0	

*For these values of t the feasibility check—Theorem 3.3.2—was violated and the procedure was started to compute the transition point $\bar{t}$.

$$P(t): \quad \min\{-tx_1x_2x_4 + (1-t)[(x_1-1.5)^2 + x_2^2 + (x_3-1.5)^2 + (x_4-1)^2 + x_5^2]$$
$$|g_1 = x_1 + x_2 + x_3 - 3 = 0, g_2 = x_1^2 + x_2^2 - x_3^2 - 2x_1x_2x_5 = 0,$$
$$g_3 = x_4^2 + x_5^2 - 1 = 0\}, \qquad [t_A, t_B] = [0, 1].$$

Geometrically, the area of a triangle of a given circumference $c = 3$ is maximized (Figure 3.3 and Table 3.2). The starting position describes a triangle of degenerate type (area = 0).

Example 3.4.3

(See Hackl [96], where this example was discussed under the aspect of ε feasibility and solved by embedding.)

$$P(t): \quad \min\{-tx_1 + (1-t)[(x_1-10)^2 + (x_2+10)^2]|g_1 = -x_2 - 10(1-t)^3 \leqslant 0,$$
$$g_2 = x_1^3 + x_2 - 990(1-t^3) \leqslant 0\}, \qquad [t_A, t_B] = [0, 1].$$

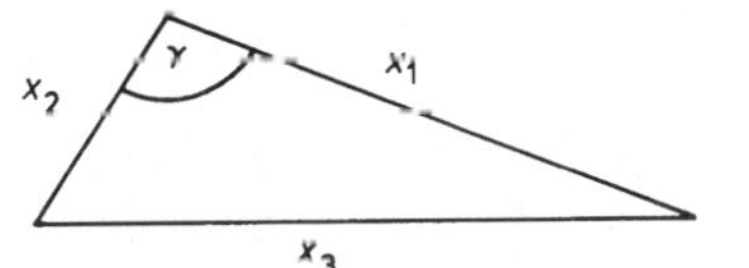

Figure 3.3

Table 3.2

t	x_1	x_2	x_3	x_4	x_5	$J(t)$
0.0	1.5	0.0	1.5	1.0	0.0	$\{1, 2, 3\}$
0.1	1.4728	0.0535	1.4737	1.0	0.0001	
0.2	1.4404	0.1147	1.4448	1.0000	0.0011	
0.3	1.4019	0.1849	1.4132	1.0000	0.0045	
0.4	1.3464	0.2649	1.3787	1.0000	0.0130	
0.5	1.3033	0.3563	1.3404	1.0000	0.0309	
0.6	1.2425	0.4603	1.2973	0.9980	0.0636	
0.7	1.1753	0.5779	1.2468	0.9930	0.1186	
0.8	1.1057	0.7091	1.1852	0.9790	0.2045	
0.9	1.0419	0.8514	1.1067	0.9942	0.3304	
1.0*	1.0000	1.0000	1.0000	0.8660	0.5000	

*Exact solution for $t = 1$: $(\bar{x}_1, \bar{x}_2, \bar{x}_3, \bar{x}_4, \bar{x}_5) = (1.0, 1.0, 1.0, \sqrt{3}/2, 0.5)$.

The feasible set for $t = 1$ is shown in Figure 3.4. The algorithm was stopped at $t = 1 - 10^{-10}$), since $\mu_j(t) \to +\infty$ as $t \to 1$ (Table 3.3). Hence, we cannot find the global minimizer $(0,0)$ for $P(1)$ by the algorithm PATH I.

Example 3.4.4

$$P(t): \quad \min\left\{ \tfrac{1}{2} \sum_{i=1}^{N} x_i^2 \,\middle|\, h_1 = x_1 + 4t - 4 = 0, g_j = -(x_{j+1} - 2)^2 - \sum_{\substack{i=2 \\ i \neq j+1}}^{N} x_i^2 \right.$$

$$+ \left(x_{j+1} - \frac{1}{N} \right)^2 - 1 \leqslant 0, j = 1, \ldots, N-1,$$

$$g_N = -\left[x_1 - 2 - \left(1 - \frac{1}{N^2} \right)^{1/2} \right]^2 + \sum_{i=2}^{N} x_i \leqslant 0 \right\}.$$

This example (Figure 3.5 and Table 3.4) allows us to check the algorithm under three aspects:

(i) Dimensionality.

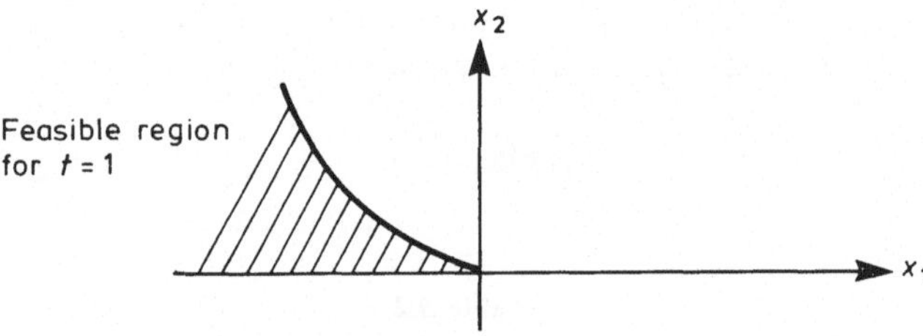

Figure 3.4

Table 3.3

t	x_1	x_2	μ_1	μ_2	$J(t)$
0	10.	$-10.$	0.0000	0.0000	
0.3	7.0000	-3.4300	9.2229	0.0306	
0.6	4.0000	-0.6400	7.6001	0.1125	
0.9	1.0000	-0.0100	2.8980	0.3000	
0.925	0.7500	-0.0042	2.8697	1.3704	Only parts of the
0.9578	0.4219	-0.0008	4.1511	3.3075	the iterations
0.9867	0.1335	-0.0000	23.6527	23.3857	are listed
$1 - 10^{-10}$	0.0000	0.0000	0.3276×10^{17}	0.3276×10^{17}	

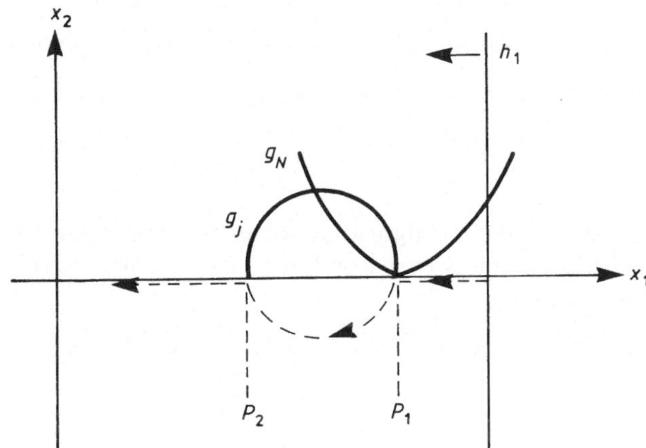

Figure 3.5

Table 3.4

t	x	$x_2=x_3=x_4$	λ_1	$\mu_1=\mu_2=\mu_3$	$J(t)$	
0.0	4.0	0.0	-4.0	0.0	$\{\varnothing\}$	
0.1	3.6	0.0	-3.6	0.0		
0.2	3.2	0.0	-3.2	0.0		
0.3	2.8	0.0	-2.8	0.0		$g_j \leqslant 0, j=1,2,3,$ violated
0.2550	2.9860	-0.0166	-2.7801	0.0333	$\{1,2,3\}$	Transition point
0.3560	2.5760	-0.3958	-4.0	0.1423		
0.4560	2.1760	-0.4931	-2.0210	0.1468		
0.5560	1.7760	-0.4881	-1.9731	0.1466		
0.6560	1.3760	-0.3745	-1.9054	0.1414		
0.6810	1.2760	-0.3203	-1.8739	0.1376		
0.7310	1.0760	-0.1330	-1.6914	0.1110		Step size control activated!
0.7451	1.0198	-0.0023	-0.9847	-0.0060		$g_j \leqslant 0, j=1,2,3,$ violated
0.7449	1.0202	-0.0001	-1.0213	0.002	$\{\varnothing\}$	Transition point
0.8459	0.6162	0.0	-0.6162	0.0		
0.9459	0.0000	0.0	-0.2162	0.0		
1.0		0.0	-0.0000	0.0		

(ii) At P_1, P_2 there is an active index set jump each of order $N - 2$.

(iii) At P_1 the constraint g_N becomes active only for the point P_1.

For $N = 5$: we have $x_5(t) \equiv 0$, $x_2 = x_3 = x_4$ and $\mu_1 = \mu_2 = \mu_3$, $[t_A, t_B] = [0, 1]$.

Example 3.4.5

(See Watson [224].) Watson's algorithm solves the nonlinear complementary problem by embedding and is based on Sard's lemma. We restrict ourselves to the following problem:

(P): $\qquad \min\left\{ f_0(x):= \exp\left(\sum_{i=1}^{5} (x_i - i + 2)^2 \right) \Big| - x_i \leq 0, i = 1, \ldots, 5 \right\}.$

Watson then solves

$$\rho_a(t, x):= tH(x) + (1 - t)(x - a) = 0$$

with

$$H(x):= -G(x) \qquad \text{and} \qquad G_i(x):= |F_i(x) - x_i|^3 - [F_i(x)]^3 - x_i^3,$$

where

$$F(x):= Df_0(x).$$

To obtain the solution $\bar{x} = (0, 0, 1, 2, 3)^T$ Watson's homotopy method requires 377 Jacobian evaluations. We used the following embedding (Table 3.5):

$$P(t): \quad \min\left\{ t^2 \exp\left(\sum_{i=1}^{5} (x_i - t + 2)^2 \right) + 10^3(1 - t^2) \sum_{i=1}^{5} x_i^2 \Big| - x_i \leq 0, i = 1, \ldots, 5 \right\},$$

$$t \in [0, 1].$$

Table 3.5*

t	x_3	x_5	$\dot{x}_5$	$J(t)$
0.0	0.0	0.0	0.0	$\{1, 2, 3, 4, 5\}$
0.0000	0.0	0.0	0.0000	Transition point: 0
0.2501	0.2238	0.6715	0.1488×10^1	
0.3001	0.2793	0.8387	0.8737	
0.4501	0.3177	0.9536	0.7050	
0.6001	0.3523	0.1057×10^1	0.6939	
0.7501	0.3900	0.1170×10^1	0.8594	
0.9001	0.4489	0.1347×10^1	0.1803×10^1	
1.0000	0.1000×10^1	0.3000×10^1	0.2207×10^4	

*Total number of iterations ($=$ Jacobian evaluations) $= 39$.

4

Pathfollowing Along a Connected Component in the Karush–Kuhn–Tucker Set and in the Critical Set

4.1 PRELIMINARY OUTLINE

The discussion of Chapter 2 has led to two generic classes of one-parametric optimization problem

$$P(t) := \min\{f(x,t)\,|\,x \in M(t)\}, \qquad t \in \mathbb{R}, \tag{4.1.1}$$

where

$$M(t) := \{x \in \mathbb{R}^n \,|\, h_i(x,t) = 0, i \in I, g_j(x,t) \leqslant 0, j \in J\},$$
$$I := \{1,\ldots,m\}, \qquad J := \{1,\ldots,s\}, \qquad m < n. \tag{4.1.2}$$

The first class is characterized by the assumption that $(f, H, G) \in \mathscr{F}^{**}$, where $\mathscr{F}^{**}$ is C_s^3 open and dense in $C^3(\mathbb{R}^n \times \mathbb{R}, \mathbb{R})^{1+m+s}$ (cf. Section 2.5). The second class contains all problems for which zero is a regular value of the associated PC^1 mapping $\mathscr{H}$ defined by

$$\mathscr{H} : \begin{pmatrix} x \\ \lambda \\ \mu \\ t \end{pmatrix} \mapsto \begin{pmatrix} D_x^{\mathrm{T}} f(x,t) + \sum_{i=1}^{m} \lambda_i D_x^{\mathrm{T}} h_i(x,t) + \sum_{j=1}^{s} \mu_j^+ D_x g_j(x,t) \\ -h_i(x,t), i = 1,\ldots,m \\ \mu_j^- - g_j(x,t), j = 1,\ldots,s \end{pmatrix} \tag{4.1.3}$$

(cf. Section 2.6).

The functions $f, h_i, g_j, i \in I, j \in J$, are assumed to belong to $C^k(\mathbb{R}^n \times \mathbb{R}, \mathbb{R})$, k equal to 2 or 3 (this will be announced later on). In comparison with the set $\mathscr{F}^{**}$ we note that $(f, H, G) \in \mathscr{F}^{**}$ implies zero to be a regular value of $\mathscr{H}$.

In Section 4.2 we consider the set $\mathscr{H}^{-1}(0)$ and assume that 0 is a regular value of $\mathscr{H}$. We describe an active index set strategy by which a connected component of the set $\mathscr{H}^{-1}(0)$ can be followed numerically. Here, a connected component is a path or a loop (cf. Theorem 2.6.1). For further information we refer to Section 2.6. Numerical results of the associated computer program written by Mörwald [166] are included in Section 4.2. Since a path in the set Σ_{stat} of stationary points can end at a point $\bar{z}$ although there exists a continuation in Σ_{gc} (cf. Remark 4.3.1 and Figures 4.2(h), (i), (k) and (l)) we walk along a connected component in Σ_{gc} in Section 4.4. Here, we restrict ourselves to the class $\mathscr{F}^{**}$ because the behaviour of all singularities that may appear is completely known (cf. Section 2.5). We note that we can use the same technique as described in Section 4.2, but we have to use additional information if a point of type 2 or type 5 occurs (bifurcation in Σ_{gc}).

4.2 PATHFOLLOWING IN THE KARUSH–KUHN–TUCKER SET

Throughout this section we assume that zero is a regular value of the mapping $\mathscr{H}$ defined by (4.1.3). Of course, then the functions $f, h_i, g_j, i \in I, j \in J$, are assumed to be in $C^2(\mathbb{R}^n \times \mathbb{R}, \mathbb{R})$. We suppose the properties of $\mathscr{H}^{-1}(0)$ and the set Σ_{stat} of all stationary points described in Section 2.6 to be known. In this section we follow Gfrerer *et al.* [71].

Let S be a connected component of $\mathscr{H}^{-1}(0)$. Since S is a path or a loop (cf. Theorem 2.6.1), each compact connected subset of S can be written in the form

$$\{\theta(s) = (x(s), \lambda(s), \mu(s), t(s)) \in \mathbb{R}^n \times \mathbb{R}^m \times \mathbb{R}^s \times \mathbb{R} \mid s \in [a,b]\},$$

where θ is a continuous function on the interval $[a,b]$. For the active index set $J_0(x(s), t(s))$ at $(x(s), t(s)) \in \mathbb{R}^n \times \mathbb{R}$ we simply write $J_0(s)$. The following theorem characterizes the possible changes of the set $J_0(s)$ if s varies in $[a,b]$.

THEOREM 4.2.1 Let zero be a regular value of $\mathscr{H}$. Then there exists a unique subdivision of $[a,b]$ into a finite number of non-trivial intervals

$$[a,b] = \bigcup_{k=1}^{K} [s^{k-1}, s^k]$$

and a corresponding family of index sets J^k, $k = 1, \ldots, K$, such that:

(i) J^k and J^{k+1} differ by exactly one index, $k = 1, \ldots, K$;

(ii) $\theta((s^{k-1}, s^k)) \subset \operatorname{int} \tau(J^k)$ (cf. Section 2.4);

(iii) $J_0(s) = J^k$ for all $s \in (s^{k-1}, s^k)$.

This theorem is a simple consequence of the transversality property. Whenever

$(x(s), \lambda(s), \mu(s))$ meets the boundary of some cell $\tau(\tilde{J})$, $\tilde{J} \subset J$ (cf. (2.4.18)), exactly one component of μ changes its sign. The intervals $[t(s^{k-1}), t(s^k)]$ are called local *stability sets*, and the points $t(s^k)$, $k = 1, \ldots, K-1$, are called *transition points*. Different stability sets may correspond to the same index set J^k. Property (i) of Theorem 4.2.1 excludes only the equality of index sets corresponding to successive stability sets.

On stability sets, Theorem 4.2.1 allows one to replace $P(t)$ by a reduced problem of the form

$$P^{\tilde{J}}(t): \qquad \min\{f(x,t)\,|\,h_i(x,t) = 0, i \in I, g_j(x,t) = 0, j \in \tilde{J}\}, \qquad \text{with } \tilde{J} \subset J.$$

The corresponding KKT system is given by

$$\mathscr{H}^{\tilde{J}}(x, u, v, t) = 0, \tag{4.2.1}$$

where $\mathscr{H}^{\tilde{J}}: \mathbb{R}^n \times \mathbb{R}^I \times \mathbb{R}^{\tilde{J}} \times \mathbb{R} \to \mathbb{R}^n \times \mathbb{R}^I \times \mathbb{R}^{\tilde{J}}$ is defined by

$$\mathscr{H}^{\tilde{J}}: \begin{pmatrix} x \\ u \\ v \\ t \end{pmatrix} \mapsto \begin{pmatrix} D_x^T f(x,t) + \sum_{i=1}^{m} u_i D_x^T h_i(x,t) + \sum_{j \in J} v_j D_x^T g_j(x,t) \\ h_i(x,t), i \in I \\ g_j(x,t), j \in \tilde{J} \end{pmatrix}.$$

The next theorem gives a computationally important characterization of stability sets.

THEOREM 4.2.2   Assume the notations and hypotheses of Theorem 4.2.1. Then, for each $k \in \{1, \ldots, K\}$, there exists an open neighbourhood $U^k \supset [s^{k-1}, s^k]$ and a continuously differentiable function $\theta^k: U^k \to \mathbb{R}^n \times \mathbb{R}^{|I|} \times \mathbb{R}^{|J^k|} \times \mathbb{R}$, $\theta(s) = (x^k(s), u^k(s), v^k(s), t(s))$, with

(i) $x^k(s) = x(s)$, $u^k(s) = \lambda(s)$, $v_j^k(s) = \mu_j(s)$, $j \in J^k$ and $t^k(s) = t(s)$ for each $s \in [s^{k-1}, s^k]$;

(ii) $\theta^k(s)$ solves (4.2.1) with $\tilde{J} = J^k$ for all $s \in U^k$;

(iii) $[s^{k-1}, s^k]$ is a connected component of $\{s \in U^k \cap [a,b]\,|\,g_j(x^k(s), t^k(s)) \leqslant 0, j \in J \setminus J^k$ and $v_j^k(s) \geqslant 0, j \in J^k\}$.

The first two properties are obtained by applying the implicit function theorem to the reduced problem. Property (iii) shows that the stability set is recognized by working with the reduced problem and by additionally computing $g_j(x^k(s), t(s))$ for $j \in J \setminus J^k$ and $v_j^k(s)/0$ for $j \in J^k$, respectively. This immediately leads to

THEOREM 4.2.3   Assume the notations and hypotheses of Theorem 4.2.1. Then, for each $k \in \{1, \ldots, K\}$, there is a unique index $j^k \in J$ with either $v^k(s^k) = 0$ and

$j^k \in J^k$ or $g_{j^k}(x^k(s^k), t^k(s^k)) = 0$ and $j^k \in J \setminus J^k$, and we have

$$J^{k+1} = \begin{cases} J^k \cup \{j^k\} & \text{if } j^k \in J \setminus J^k, \\ J^k \setminus \{j^k\} & \text{if } j^k \in J^k. \end{cases}$$

This theorem describes how the index set J^{k+1} differs from the index set J^k.

In the following we describe the pathfollowing method on U^k (cf. Theorem 4.2.2).

On a local stability set $[s^{k-1}, s^k]$ the KKT conditions for $P(t)$ reduce to a system of equations $\mathcal{H}^J(x, u, v, t) = 0$ with $\bar{J} = J^k$. Pathfollowing algorithms for such systems are well known (cf. e.g. the references in Section 1.1).

Let $\pi^{p-1} \in \{-1, +1\}$ and $J^{p-1} \subset J$. We assume that $w^{p-1} = (x^{p-1}, u^{p-1}, v^{p-1}, t^{p-1}) \in \mathbb{R}^n \times \mathbb{R}^{|I|} \times \mathbb{R}^{|J^{p-1}|} \times \mathbb{R}$ approximates a solution of $\mathcal{H}^{J^{p-1}}(x, u, v, t) = 0$ with $g_j(x^{p-1}, t^{p-1}) \leqslant 0$ for all $j \in J \setminus J^{p-1}$ and $v_j^{p-1} \geqslant 0$ for all $j \in J^{p-1}$.

Then, a typical iteration step consists of three parts. In order to proceed along S we first take a predictor step.

(a) *Predictor step* for the system $\mathcal{H}^{J^{p-1}}(w) = 0$. From w^{p-1} we take a first step along a direction y^{p-1} with the step length σ^{p-1}:

$$w^{p,0} = w^{p-1} + \sigma^{p-1} \cdot y^{p-1},$$

where y^{p-1} solves the problem

$$A^{p-1} y = 0, \qquad \|y\| = 1, \qquad \text{sign det} \begin{bmatrix} A^{p-1} \\ y^{\mathrm{T}} \end{bmatrix} = \pi^{p-1}$$

for some approximation A^{p-1} of the Jacobian $D\mathcal{H}^{J^{p-1}}(w^{p-1})$ (y^{p-1} is an approximation of the unit tangent vector of the solution path near w^{p-1}; cf. Figure 4.1). Next we use $w^{p,0}$ as the starting point of an iteration.

(b) *Corrector step* for the system $\mathcal{H}^{J^{p-1}}(w) = 0$:

$$A^{p,q-1}(w^{p,q} - w^{p,q-1}) + \mathcal{H}^{J^{p-1}}(w^{p,q-1}) = 0,$$

$$(y^{p,q-1})^{\mathrm{T}}(w^{p,q} - w^{p,q-1}) = 0 \qquad q = 1, \dots, q_p,$$

where $A^{p,q-1}$ approximates $D\mathcal{H}^{J^{p-1}}(w^{p,q})$. Usually, $y^{p,q-1}$ is equal to y^{p-1} or to one of the natural basis vectors of $\mathbb{R}^n \times \mathbb{R}^{|I|} \times \mathbb{R}^{|J^{p-1}|} \times \mathbb{R}$.

Then we set

$$\bar{w}^p = (\bar{x}^p, \bar{u}^p, \bar{v}^p, t^p) = w^{p,q_p}$$

for the next approximation of a solution of $\mathcal{H}^{J^{p-1}}(w) = 0$.

This general class of predictor–corrector methods includes the Euler predictor ($A^{p-1} := D\mathcal{H}^{J^{p-1}}(w^{p-1})$), the Newton corrector ($A^{p,q-1} = D\mathcal{H}^{J^{p-1}}(w^{p,q})$; $y^{p,q} = y^p$) and also quasi-Newton techniques.

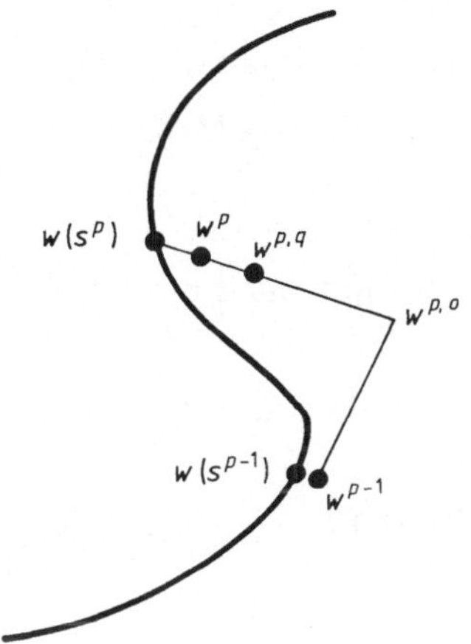

Figure 4.1

Figure 4.1 illustrates the predictor–corrector procedure. In this figure it is assumed that $y^{p,q} = y^p$, $q = 1, \ldots, q_p$.

Before finally accepting $\bar{w}^p$ as the next iterated point w^p we have to make a check.

(c) *Check* whether the local stability set was left:
If

$$g_j(\bar{x}^p, \bar{t}^p) \leq 0 \text{ for all } j \in J \setminus J^{p-1}, \ v_j^p \geq 0 \text{ for all } j \in J^{p-1}, \qquad (4.2.2)$$

then set

$$w^p = \bar{w}^p, \ J^p = J^{p-1}, \ \pi^p = \pi^{p-1}.$$

(See Theorem 4.2.2; the index set is not changed because the local stability set is not left. To keep π constant means that the component S is always followed in the same direction.)

If the boundary point s^k of the stability set $[s^{k-1}, s^k]$ is passed, the index set $J_0(s)$ changes from J^{p-1} to J^p; see Theorem 4.2.1. Therefore, if there is an index $j^p \in J$ with either

$$g_{j^p}(\bar{x}^p, \bar{t}^p) > 0 \text{ and } j^p \in J \setminus J^{p-1} \qquad \text{or} \qquad \bar{u}_{j^p}^p < 0 \text{ and } j^p \in J^{p-1}, \quad (4.2.3)$$

then change the index set (cf. Theorem 4.2.3):

$$J^p = \begin{cases} J^{p-1} \cup \{j^p\} & \text{if } j^p \in J \setminus J^{p-1}, \\ J^{p-1} \setminus \{j^p\} & \text{if } j^p \in J^{p-1}, \end{cases} \qquad (4.2.4)$$

and set

$$w^p = (\bar{x}^p, \bar{u}^p, \bar{v}^p, \bar{t}^p) \in \mathbb{R}^n \times \mathbb{R}^{|I|} \times \mathbb{R}^{|J^p|} \times \mathbb{R}$$

with

$$v_j^p = \begin{cases} \bar{v}_j^p & \text{if } j \in J^p \cap J^{p-1}, \\ 0 & \text{if } j \in J^p \setminus J^{p-1}, \end{cases}$$

and

$$\pi^p = \pi^{p-1}.$$

The change of the sign of π is motivated by

LEMMA 4.2.1   Assume the notations and hypotheses of Theorem 4.2.1. Then, for $k \in \{1, \ldots, K-1\}$, we have

$$\lim_{s \to s^k-} \text{sign} \det \begin{bmatrix} D\mathcal{H}^{J^k}(\theta^k(s)) \\ \dot{\theta}^k(s) \end{bmatrix} = - \lim_{s \to s^k+} \text{sign} \det \begin{bmatrix} D\mathcal{H}^{J^{k+1}}(\theta^{k+1}(s)) \\ \dot{\theta}^{k+1}(s) \end{bmatrix}.$$

Proof   In the book by Garcia and Zangwill [62, Section 2.3]

$$\dot{\theta}_j(s) = c(s) \det \begin{bmatrix} (e^j)^{\mathrm{T}} \\ \hdashline D\mathcal{H}(\theta(s)) \end{bmatrix} \qquad \text{for all } s \in [a,b] \setminus \{s^0, s^1, \ldots, s^k\}$$

is shown to hold, where sign $c(s)$ is constant with respect to s, and e^j is the jth natural basis vector in $\mathbb{R}^n \times \mathbb{R}^m \times \mathbb{R}^s \times \mathbb{R}$. Then, by deleting all columns with $j \in J \setminus J^k$ and the corresponding rows, we obtain

$$\det \begin{bmatrix} (e^j)^{\mathrm{T}} \\ \hdashline D\mathcal{H}(\theta(s)) \end{bmatrix} = (-1)^{|J \setminus J^k|} \cdot \det \begin{bmatrix} (e^j)^{\mathrm{T}} \\ \hdashline D\mathcal{H}^{J^k}(\theta^k(s)) \end{bmatrix}.$$

Here e^j denotes the jth natural basis vector in $\mathbb{R}^n \times \mathbb{R}^m \times \mathbb{R}^s \times \mathbb{R}$ and $\mathbb{R}^n \times \mathbb{R}^m \times \mathbb{R}^{|J^k|} \times \mathbb{R}$, respectively. Now,

$$\det \begin{bmatrix} (\dot{\theta}^k(s))^{\mathrm{T}} \\ \hdashline D\mathcal{H}^{J^k}(\theta^k(s)) \end{bmatrix} = \sum_j \dot{\theta}_j^k(s) \cdot \det \begin{bmatrix} (e^j)^{\mathrm{T}} \\ \hdashline D\mathcal{H}^{J^k}(\theta^k(s)) \end{bmatrix}$$

$$= c(s) \cdot (-1)^{|J \setminus J^k|} \cdot \sum_j \det^2 \begin{bmatrix} (e^j)^{\mathrm{T}} \\ \hdashline D\mathcal{H}^{J^k}(\theta^k(s)) \end{bmatrix}.$$

Since the cardinality of J^k differs by 1, the proof is complete. $\square$

4.3   THE ALGORITHM PATH II AND NUMERICAL RESULTS

Algorithm PATH II

We assume a KKT point $w^0 = (x^0, \lambda^0, \mu^0, t(a))$ to be known and belongs to the connected component S. Let $\pi^{p-1} \in \{-1, +1\}$ and $J^{p-1} \subset J$. Assume that $w^{p-1} = (x^{p-1}, u^{p-1}, v^{p-1}, t^{p-1}) \in \mathbb{R}^n \times \mathbb{R}^{|I|} \times \mathbb{R}^{|J^{p-1}|} \times \mathbb{R}$ approximates a solution of $\mathcal{H}^{J^{p-1}}(x, u, v, t) = 0$ with $g_j(x^{p-1}, t^{p-1}) \leqslant 0$ for all $j \in J \setminus J^{p-1}$ and $v_j^{p-1} \geqslant 0$ for all $j \in J^{p-1}$. Then the pth iteration step is of the following form.

(a) *Predictor* step for the system $\mathcal{H}^{J^{p-1}}(w) = 0$:

$$w^{p,0} = w^{p-1} + \sigma^{p-1} \cdot y^{p-1},$$

where y^{p-1} solves

$$A^{p-1} y = 0 \qquad \|y\| = 1, \qquad \operatorname{sign} \det \begin{bmatrix} A^{p-1} \\ y^{\mathrm{T}} \end{bmatrix} = \pi^{p-1}.$$

(b) *Corrector* step for the system $\mathcal{H}^{J^{p-1}}(w) = 0$:

$$A^{p,q-1}(w^{p,q} - w^{p,q-1}) + \mathcal{H}^{J^{p-1}}(w^{p,q-1}) = 0, \qquad q = 1, \ldots, q_p,$$
$$(y^{p,q-1})(w^{p,q} - w^{p,q-1}) = 0, \qquad q = 1, \ldots, q_p.$$

Set

$$\bar{w}^p = (\bar{x}^p, \bar{u}^p, \bar{v}^p, \bar{t}^p) = \bar{w}^{p,q_p}.$$

(c) *Check* whether the local stability set was left:

If $g_j(\bar{x}^p, \bar{t}^p) \leqslant 0$ for all $j \in J \setminus J^{p-1}$, $\bar{v}_j^p \geqslant 0$ for all $j \in J^{p-1}$, then set

$$w^p = \bar{w}^p, \qquad J^p = J^{p-1}, \qquad \pi^p = \pi^{p-1}.$$

(d) *Otherwise*, there is an index $j^p \in J$ with either $g_{j^p}(\bar{x}^p, \bar{t}^p) > 0$ and $j^p \in J \setminus J^{p-1}$ or $\bar{v}_{j^p}^p < 0$ and $j^p \in J^p$, then change the index set

$$J^p = \begin{cases} J^{p-1} \cup \{j^p\} & \text{if } j^p \in J \setminus J^{p-1}, \\ J^{p-1} \setminus \{j^p\} & \text{if } j^p \in J^{p-1}, \end{cases}$$

and set

$$w^p = (\bar{x}^p, \bar{u}^p, \bar{v}^p, \bar{t}^p) \in \mathbb{R}^n \times \mathbb{R}^{|I|} \times \mathbb{R}^{|J^p|} \times \mathbb{R}$$

with

$$v_j^p = \begin{cases} \bar{v}_j^p & \text{if } j \in J^p \cap J^{p-1}, \\ 0 & \text{if } j \in J^p \setminus J^{p-1}, \end{cases}$$

and

$$\pi^p = -\pi^{p-1}.$$

From the above consideration the next theorem follows.

THEOREM 4.3.1   Let zero be a regular value of $\mathcal{H}$. Then, the algorithm PATH II, starting at $(x^a, \lambda^a, \mu^a, t(a))$, generates a discretization

$$t(a) = t_0 < \cdots < t_i < t_{i+1} < \cdots < t_N = t(b)$$

and corresponding points $(\tilde{x}^i, \tilde{\lambda}^i, \tilde{\mu}^i)$, $i = 1, \ldots, N$, with $\|(\tilde{x}^i, \tilde{\lambda}^i, \tilde{\mu}^i) - (x(t_i),$

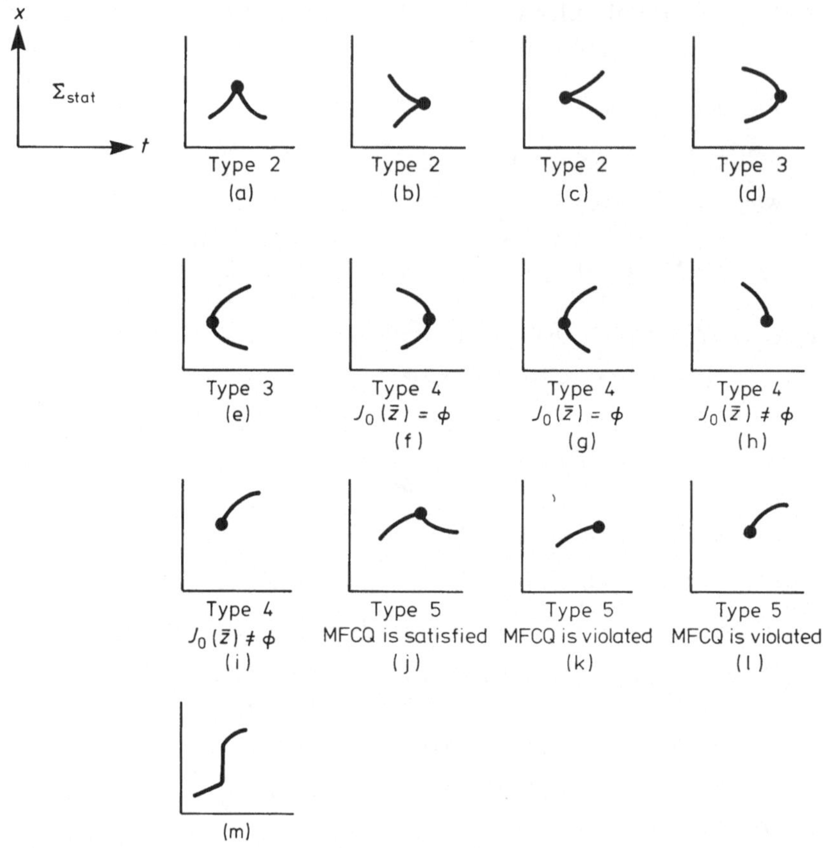

Figure 4.2

$\lambda(t_i), \mu(t_i) \| < \varepsilon$ ($\varepsilon > 0$ sufficiently small) in a finite number of steps, where $(x(t_i), \lambda(t_i), \mu(t_i)) \in S$ (S is the considered connected component in $\mathcal{H}^{-1}(0)$).

REMARK 4.3.1   (i) Since $(f, H, G) \in \mathcal{F}^{**}$ implies that zero is a regular value of $\mathcal{H}$, the cases (a)–(e) depicted in Figure 4.2 are included. We note that a Lagrangian multiplier $v_p \in \{\lambda_i, i \in I, \mu_j, j \in J_0(\bar{z})\}$ corresponding to a point $\bar{z}$ of type 5, where MFCQ is not satisfied (Figures 4.2(k) and (l)), could be the behaviour depicted in Figure 4.3 (cf. Example 4.3.1; in particular Figure 4.1). Example 4.3.1 also contains two turning points of type 2. The situation illustrated in Figures 4.2(h), (i), (k) and (l) clearly shows the disadvantage of walking along a connected component in Σ_{stat}: the path stops at $\bar{z}$, but there is a continuation in Σ_{gc} (cf. Figure 2.16).

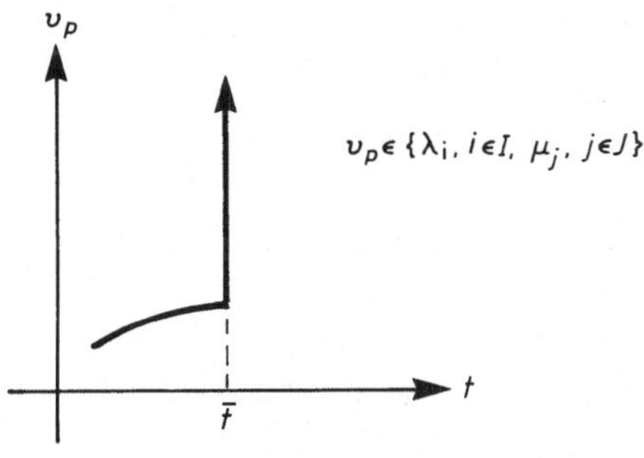

Figure 4.3

(ii) Furthermore, the behaviour of the stationary point depicted in Figure 4.2(m) is possible (cf. Section 2.6).

REMARK 4.3.2 Using the results and the algorithm PATH I of Chapter 3, it is possible to consider also the change of more than one index at transition points not being turning points (cf. Example 4.3.2).

The computer program written by Mörwald [166] includes the algorithm

Table 4.1

t	x_1	x_2	μ_1	μ_2	μ_3	μ_4	ACTIVE INDEX SET
0.0	0.0	0.0	2.0	2.0			$\{1,2\}$
0.25	0.0	0.0	2.0	2.0			
0.5	0.0	0.0	2.0	2.0			
0.5	0.0	0.0	2.0	2.0	0.0		$\{1,2,3\}$
0.5	0.0	0.0	1.0	4.0	1.0		
0.5	0.0	0.0	0.0	6.0	2.0		
0.5	0.0	0.0		6.0	2.0		$\{2,3\}$
0.625	0.5	0.0		4.0	1.0		
0.75	1.0	0.0		2.0	0.0		
0.75	1.0	0.0		2.0			$\{2\}$
0.2	1.0	0.0		2.0			
0.2	1.0	0.0		2.0		0.0	$\{2,4\}$
0.6	2.0	0.0		3.0		1.0	
1.0	3.0	0.0		4.0		2.0	

100   *Parametric Optimization: Singularities, Pathfollowing and Jumps*

PATH II and implements also the change of more than one index at points not being turning points (cf. Remark 4.3.2).

Example 4.3.1

$P(t)$: $\min\{f(x,t)\,|\,g_j(x,t)\leqslant 0, j=1,\ldots,4\}$, $t\in[0,1]$,

where

$$f(x,t):=-(x_1-1)^2-(x_2-1)^2,$$
$$g_1(x,t):=-x_1,$$
$$g_2(x,t):=-x_2,$$
$$g_3(x,t):=-x+2x_2-2+4t\leqslant 0,$$
$$g_4(x,t):=2x_1+x_2-1-5t\leqslant 0.$$

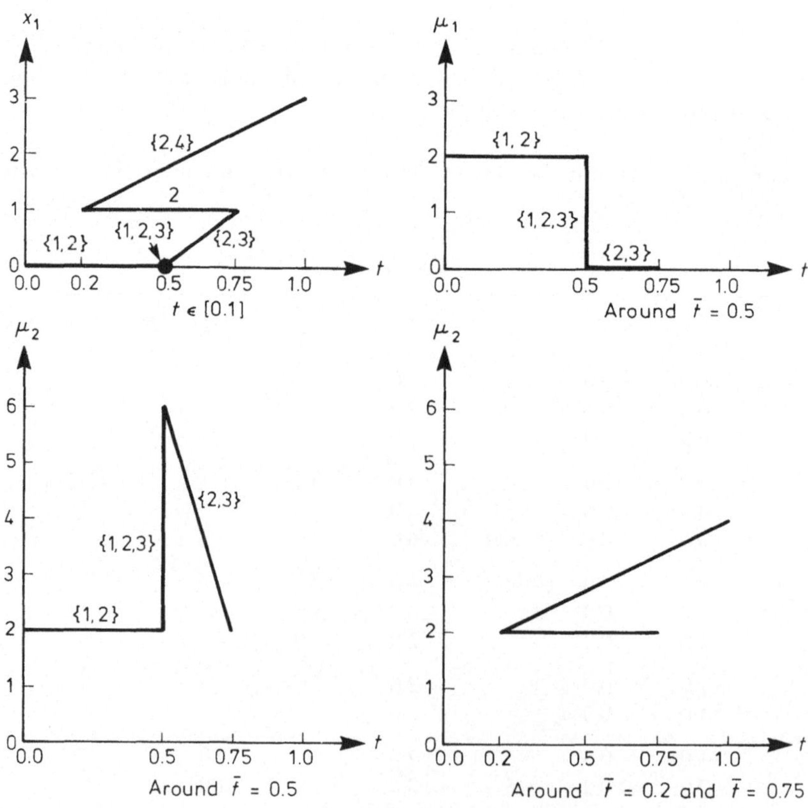

Figure 4.4

We see that this example (Table 4.1) changes the active index set four times, and exactly one index is changed by the formulae (4.16). Furthermore, we consider the behaviour of $x_1(t)$ for $t \in [0, 1]$ ($x_2(t) = 0$ for all $t \in [0, 1]$), $\mu_1(t)$ and $\mu_2(t)$ around $\bar{t} = 0.5$ and $\mu_2(t)$ around $\bar{t} = 0.2$ and $\bar{t} = 0.75$ in Figure 4.4.

We have turning points for $\bar{t} = 0.75$ and $\bar{t} = 0.2$ at which the active index set is changed by exactly one index. Furthermore, we see that for $\bar{t} = 0.5$ the set of all Lagrange multipliers associated with $x(\bar{t}) = (0, 0)$ is a line segment (cf. Section 2.6) approximated by the algorithm PATH II. The program needs 95 evaluations of the Jacobian. The number can be reduced to 13 if the step size is extended after the change of the indices.

Example 4.3.2

(See also Example 3.4.4.)

$$P(t): \qquad \min\{ f(x, t) \,|\, h_1(x, t) = 0, g_j(x, t) \leqslant 0, j = 1, \dots, N \}, \qquad t \in [0, 1],$$

where

$$f(x, t) := \tfrac{1}{2} \sum_{i=1}^{N} x_i^2,$$

$$h_1(x, t) := x_1 + 4t - 4,$$

$$g_j(x, t) := -(x_{j+1} - 2)^2 - \sum_{i=1}^{N} x_i^2 - \left(x_{j+1} - \frac{1}{N} \right)' - 1, \qquad j = 1, \dots, N - 1,$$

$$g_N(x, t) := -\left[x_1 - 2 - \left(1 - \frac{1}{N^2} \right)^{1/2} \right]^2 + \sum_{i=2}^{N} x_i.$$

The feasible set and the path of global minimizers are illustrated in Chapter 3.

The index set of active constraints is changed at the points P_1 and P_2. We have a change of $N - 1$ indices at P_1. N inequality constraints are active at P_2. Around this point we have a change of $N - 1$ indices.

This example (Table 4.2) is useful for testing the program on the following aspects:

(i) dimension (since $N = 2, 3, \dots$ is arbitrary),

(ii) the change of the active index set, and

(iii) the sensitivity at the transition points ($h_N(x, t) \leqslant 0$ is active only at the point P_1).

For $N = 4$, the program finds the solution in Table 4.2 on the IBM 4341 computer with 77 evaluations of the Jacobians after 6 s.

By increasing N, the computing time increases as follows: for $N = 40$ (the

Table 4.2

t	x_1	$x_2 = x_3 = x_4$	λ_1	$\mu_1 = \mu_2 = \mu_3$	ACTIVE INDEX SET
0.00000	4.00000	0.00000	-4.00000		$\{\varnothing\}$
0.25050	2.99796	-0.00000	-2.99796		
0.25051	2.97979	-0.00000	-2.97979	0.00000	$\{1, 2, 3\}$
0.27954	2.88183	-0.18874	-2.23017	0.12316	
0.69831	1.20673	-0.27199	-1.84384	0.13385	
0.74499	1.02000	0.00098	-1.00529	-0.00250	
0.74494	1.02020	-0.00000	-1.02020		$\{\varnothing\}$
0.9399	0.24033	-0.00000	-0.24033		
1.00000	0.00000	-0.00000	-0.00000		

largest problem for the test), the computing time is 25 min with about 200 computations of the Jacobians.

Furthermore, we note that all examples of Chapter 3 were successfully tested. By the way, a practical example of the optimal control of a water power plant with one storage for a given time period was solved by this program (cf. Mörwald [166]). The problem has 41 constraints and 20 variables. The computing time was limited to 30 min for practical reasons. For more details we refer to [166].

4.4 PATHFOLLOWING IN THE CRITICAL SET

Throughout this section we assume that $(f, H, G) \in \mathscr{F}^{**}$. Of course, then the functions $f, h_i, g_j, i \in I, j \in J$, are assumed to be in $C^3(\mathbb{R}^n \times \mathbb{R}, \mathbb{R})$. We can weaken the assumption in some cases. We suppose the contents of Section 2.5. Here, we follow Guddat *et al.* [84].

Let $C = C(x^0, 0)$ be the connected component in Σ_{gc} containing $(x^0, 0)$, and $(x^0, 0)$ is assumed to be a point of type 1. We assume that $(x^0, 0)$ is known or easy to compute. Let $Z = Z(x^0, 0)$ be the connected component of $C \cap \Sigma^1_{gc}$ containing $(x^0, 0)$. Then, from Section 2.5 we know that the closure $\bar{Z}$ can be uniquely described by

$$\bar{Z} = \{(x, t) \in \mathbb{R}^n \times \mathbb{R} \mid (x(t), t), t \in [\underline{t}, \bar{t}]\}, \tag{4.4.1}$$

where $x: [\underline{t}, \bar{t}] \to \mathbb{R}^n$ is a function that is at least twice continuously differentiable on $(\underline{t}, \bar{t})$ and continuous on $[\underline{t}, \bar{t}]$ and $-\infty \leqslant \underline{t} < 0 < \bar{t} \leqslant +\infty$ and $x(0) = x^0$.

Of course, $z(t) = (x(t), t)$ is a point of type 1 for all $t \in (\underline{t}, \bar{t})$, i.e. $z(t)$ is a non-degenerate critical point for $P(t)$ for all $t \in (\underline{t}, \bar{t})$ (cf. Definition 2.4.3).

If $\bar{t}$ is finite, then $\bar{z} = (\bar{x}, \bar{t})$ with $x(\bar{t}) = \bar{x}$ is a point of type i, $i \in \{2, 3, 4, 5\}$. Of course, we have the same situation for $\underline{t}$ if $\underline{t}$ is finite. Throughout this section we consider only the right boundary point of $[\underline{t}, \bar{t}]$. It is easy to replace $\bar{t}$ by $\underline{t}$ in the following investigation.

Now, as in Chapter 3 and Section 4.2, we introduce the notion of a local stability set.

DEFINITION 4.4.1 $S(x^0, t_0) = [\underline{t}, \bar{t}]$ is called the *local stability set* with respect to C and the point (x^0, t_0) with $t_0 = 0$. Then $\underline{t}$ and $\bar{t}$ are called *transition points*.

Of course, this definition is also possible for an arbitrary point $(x^0, t_0) \in C \cap \Sigma^1_{gc}$.

We recall the indices LI, LCI and QI, QCI, respectively, introduced and used in Section 2.5, which represent the number of negative or positive values of the Lagrange multipliers and the number of negative or positive eigenvalues of the Hessian $D_x^2 L(x, t)|_{T_x M}$ at the point (x, t). Of course, these indices depend on the parameter t. This is the reason why we denote them by LI(t), LCI(t), QI(t), QCI(t) Using the characterization of points of type 1 (cf. Section 2.5) we get

LEMMA 4.4.1 Let $[\underline{t}, \bar{t}]$ be the local stability set (cf. Definition 4.4.1) and $x: [\underline{t}, \bar{t}] \to \mathbb{R}^n$ the function defined by (4.4.1). Then x has the following properties for all $t \in [\underline{t}, \bar{t})$:

(1)　(i)　$J_0(x(t), t) = \text{const}(=: J_0)$,

　　(ii)　$\text{LI}(t) = \text{const}(= a)$,
　　　　$\text{LCI}(t) = \text{const}(= b)$,

　　(iii)　$\text{QI}(t) = \text{const}(= c)$,
　　　　$\text{QCI}(t) = \text{const}(= d)$,

　　(iv)　$|J_0| = a + b$.

(2)　(i)　$D_x h_i(x(t), t)$, $i \in I$, $D_x g_j(x(t), t)$, $j \in J_0$, are linearly independent (LICQ),

　　(ii)　the Lagrange multiplier vector $(\lambda(t), \mu(t)) \in \mathbb{R}^m \times \mathbb{R}^{|J_0|}$ is uniquely determined and $\mu_j(t) \neq 0$, $j \in J_0$,

　　(iii)　$(\lambda, \mu) \in C^2((\underline{t}, \bar{t}), \mathbb{R}^m \times \mathbb{R}^{|J_0|})$,

　　(iv)　$D_x^2 L(x(t), t)|_{T_{x(t)} M}$ is non-singular.

REMARK 4.4.1 On the basis of this lemma we can analyse the character of the critical point, e.g. we have $a + c = 0$ for a local minimizer. Consequently, for the knowledge of c we need an estimation of the smallest eigenvalue of the matrix $D_x^2 L(x(t), t)|_{T_{x(t)} M}$. For this we refer to Richter [182]. Of course, the observation of the linear index a is very easy.

In the following we analyse the continuation of $x(t)$ at the transition point t depending on the different types $i \in \{2, 3, 4, 5\}$.

4.4.1 A point of type 2

Let $x(\bar{t}) = \bar{x}$ and $\bar{z} = (\bar{x}, \bar{t})$ be a point of type 2 and $\bar{t}$ the right boundary point of the local stability set $[\underline{t}, \bar{t}]$. Then, the behaviour of C around $\bar{z}$ is described in Figure 2.13.

REMARK 4.4.2   We see that $\bar{z} = (\bar{x}, \bar{t})$ is a bifurcation point in Σ_{gc} with two continuation of $x(t)$ in Σ_{gc}. The complementarity condition is violated at $\bar{z}$. If we restrict ourselves to Σ_{loc} as in Chapter 3 or to Σ_{stat} as in Section 4.2, we have one continuation; see Figure 4.5.

The dotted line in Figure 4.5(b) represents saddle points at which the quadratic index $c = 1$. The index c is the same around $\bar{z}$ in Figure 4.5(c), while the index differs exactly by one around $\bar{z}$ in Figure 4.5(d).

Let J_0 be the index set defined by Lemma 4.4.1. Now we consider the following problem with equality constraints only (cf. 4.2.1):

$$P^{J_0}(t): \qquad \min\{f(x,t) \mid h_i(x,t) = 0, i \in I, g_j(x,t) = 0, j \in J_0\}, \qquad t \in \mathbb{R}. \qquad (4.4.2)$$

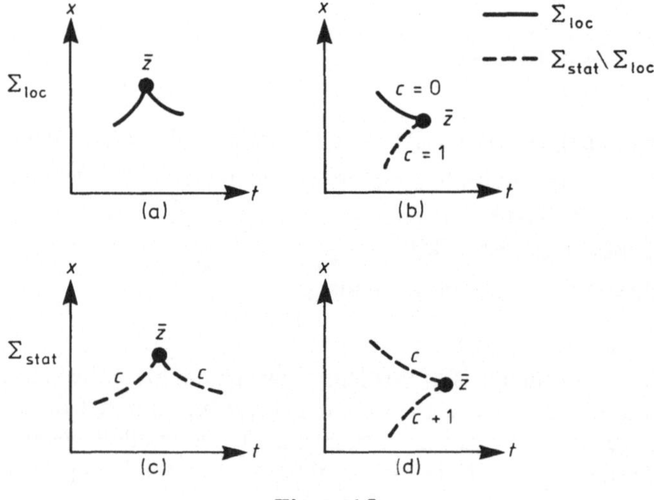

Figure 4.5

Then, the corresponding KKT system has the following description:

$$D_x f(x,t) + \sum_{i\in I} \lambda_i D_x h_i(x,t) + \sum_{j\in J_0} \mu_j D_x g_j(x,t) = 0,$$

$$h_i(x,t) = 0, \qquad i\in I,$$
$$g_j(x,t) = 0, \qquad j\in J_0, \qquad\qquad (4.4.3)$$
$$\lambda_i\in\mathbb{R}, i\in I, \qquad \mu_j\in\mathbb{R}, j\in J_0.$$

Analogously to Lemma 3.3.2 we have

LEMMA 4.4.2 Assume the notations and hypothesis of Lemma 4.4.1. Then there exists a $\delta>0$ and a function $(x^{J_0}, \lambda^{J_0}, \mu^{J_0})\in C^2((\underline{t}, \bar{t}+\delta), \mathbb{R}^n\times\mathbb{R}^m\times\mathbb{R}^{|J_0|})$ with the following properties:

(i) $(x^{J_0}(t), \lambda^{J_0}(t), \mu^{J_0}(t))$ solves (4.4.3) for all $t\in(\underline{t}, \bar{t}+\delta)$,

(ii) $x^{J_0}(t) = x(t)$, $\lambda^{J_0}(t) = \lambda(t)$, $\mu_j^{J_0}(t) = \mu_j(t)$, $j\in J_0$, for all $t\in[\underline{t},\bar{t}]$,

(iii) $D_x h_i(x^{J_0}(t), t)$, $i\in I$, $D_x g_j(x^{J_0}(t), t)$, $j\in J_0$, are linearly independent (LICQ) for all $t\in[\underline{t}, \bar{t}+\delta)$,

(iv) $D_x^2 L(x^{J_0}(t), t)|_{T_x J_{0(t)}M}$ is regular for all $t\in(\underline{t}, \bar{t}+\delta)$.

From (i) and (iii) it follows that $x^{J_0}(t)$ is a critical point for $P^{J_0}(t)$ for all $t\in(\underline{t}, \bar{t}+\delta)$ (cf. Definition 2.4.2). We can follow the curve $\{(x^{J_0}(t), \lambda^{J_0}(t), \mu^{J_0}(t)) | t\in (\underline{t}, \bar{t}+\delta)\}$ numerically by applying a modification of the standard algorithms PATH I and PATH II, respectively, for (4.4.3) with the starting point $(x^0, 0)$ for $t>0$ and $t<0$. We note that both algorithms are the same in this situation. Of course, $C = C(x^0, 0) \subset \Sigma_{gc}$ and for the numerical tracing of the fully connected component we have to follow in both directions.

First, we have to modify the check whether the local stability set was left (instead of (4.1.5)):

$$g_j(x^{J_0}(t), t)\leqslant 0 \qquad\text{for all } j\in J\backslash J_0,$$
$$\operatorname{sign}\mu_j^{J_0}(0) = \operatorname{sign}\mu_j^{J_0}(t), \qquad j\in J_0$$

(cf. Section 2.5 for a point of type 2 and Lemma 4.4.1).

Secondly, we have to find the new index sets for the two continuations in Σ_{gc} (cf. Figure 2.13). The answer is given by

THEOREM 4.4.1 Assume the notations and hypothesis of Lemma 4.4.2. Then we have the following cases of continuations at $\bar{z} = (\bar{x}, \bar{t})$.

Case I: There exists a $j_0\in J\backslash J_0$ with $g_{j_0}(x^{J_0}(\bar{t} - \varepsilon)) < 0$, $g_{j_0}(x^{J_0}(\bar{t} + \varepsilon)) > 0$ for $\varepsilon\in(0, \delta)$ sufficiently small, then it holds for $J_k = J_0\cup\{j_0\}$.

There exist $\gamma > 0$ and a unique function

$$(x^{J_k}, \lambda^{J_k}, \mu^{J_k}) \in C^2((\bar{t} - \gamma, \bar{t} + \gamma), \mathbb{R}^n \times \mathbb{R}^m \times \mathbb{R}^{|J_k|}$$

with the following properties:

(i) $(x^{J_k}(t), \lambda^{J_k}(t), \mu^{J_k}(t))$ is a solution of (4.4.3) for $J_0 = J_k$ for all $t \in (\bar{t} - \gamma, \bar{t} + \gamma)$,

(ii) $x^{J_k}(\bar{t}) = x^{J_0}(\bar{t})$, $\lambda^{J_k}(t) = \lambda^{J_0}(\bar{t})$, $\mu_j^{J_k}(\bar{t}) = \begin{cases} \mu_j^{J_0}(\bar{t}), & j \in J_0, \\ 0, & j = j_0, \end{cases}$

(iii) $x^{J_k}(t)$ is a non-degenerate critical point for $P^{J_k}(t)$ for all $t \in (\bar{t} - \gamma, \bar{t} + \gamma), t \neq \bar{t}$,

(iv) $x^{J_k}(t)$ is a critical point for $P(t)$ for all $t \in (\bar{t} - \gamma, \bar{t} + \gamma)$.

Case II: There exists a $p \in J_0$ with either

$$\mu_p^{J_0}(\bar{t} - \varepsilon) > 0, \qquad \mu_p^{J_0}(\bar{t} + \varepsilon) < 0,$$

or

$$\mu_p^{J_0}(\bar{t} - \varepsilon) < 0, \qquad \mu_p^{J_0}(\bar{t} + \varepsilon) > 0,$$

for $\varepsilon \in (0, \delta)$ sufficiently small, then we have the following continuations:

Case IIa: $J_k = J_0$ with the same orientation of t as in $P^{J_0}(t)$ and

Case IIb: $J_k = J_0 \setminus \{p\}$ with either the same or the opposite orientation of t as in $P^{J_0}(t)$, and there exist $\gamma > 0$ and a unique function

$$(x^{J_k}, \lambda^{J_k}, \mu^{J_k}) \in C^2((\bar{t} - \gamma, \bar{t} + \gamma), \mathbb{R}^n \times \mathbb{R}^m \times \mathbb{R}^{|J_k|})$$

with the following properties:

(i) $(x^{J_k}(t), \lambda^{J_k}(t), \mu^{J_k}(t))$ is a solution of (4.4.3) for $J_0 = J_k$ for all $t \in (\bar{t} - \gamma, \bar{t} + \gamma)$,

(ii) $x^{J_k}(\bar{t}) = x^{J_0}(\bar{t})$, $\lambda^{J_k}(\bar{t}) = \lambda^{J_0}(\bar{t})$, $\mu_j^{J_k}(\bar{t}) = \mu_j^{J_0}(\bar{t}), j \in J_0 \setminus \{p\}$,

(iii) $x^{J_k}(t)$ is a non-degenerate critical point for $P^{J_k}(t)$ for all $t \in (\bar{t} - \gamma, \bar{t} + \gamma)$, $t \neq \bar{t}$,

(iv) $x^{J_k}(t)$ is a critical point for $P(t)$ either for all $t \in (\bar{t}, \bar{t} + \gamma)$ or for all $t \in (\bar{t} - \gamma, \bar{t})$ depending on the orientation.

This theorem is a consequence of the results presented in Section 2.5.

In Figure 4.6 all situations possible are illustrated for the left and the right boundary points of the local stability set $[\underline{t}, \bar{t}]$ (cf. also Figure 2.13).

With respect to case IIb (cf. (iv)) we know the new index set $J_k = J_0 \setminus \{p\}$, but we do not know the orientation of t after $\bar{t}$. In this case we propose to try to solve $P^{J_k}(t)$, $t \geq \bar{t}$, t sufficiently close to $\bar{t}$. Then, if we find a critical point for $P(t)$, we will have the right orientation. In the other case we have to consider $P^{J_k}(t)$, $t \leq \bar{t}$.

x(t)
Continuations in Σ_{gc}
Case I
Case II
(a)
(b)
(c)
(d)
(e)
(f)
(g)
(h)
(i)
(j)
(k)
(l)
J_k
J_0
$\mu_p > 0$
$\mu_p < 0$
$\mu_p = 0$

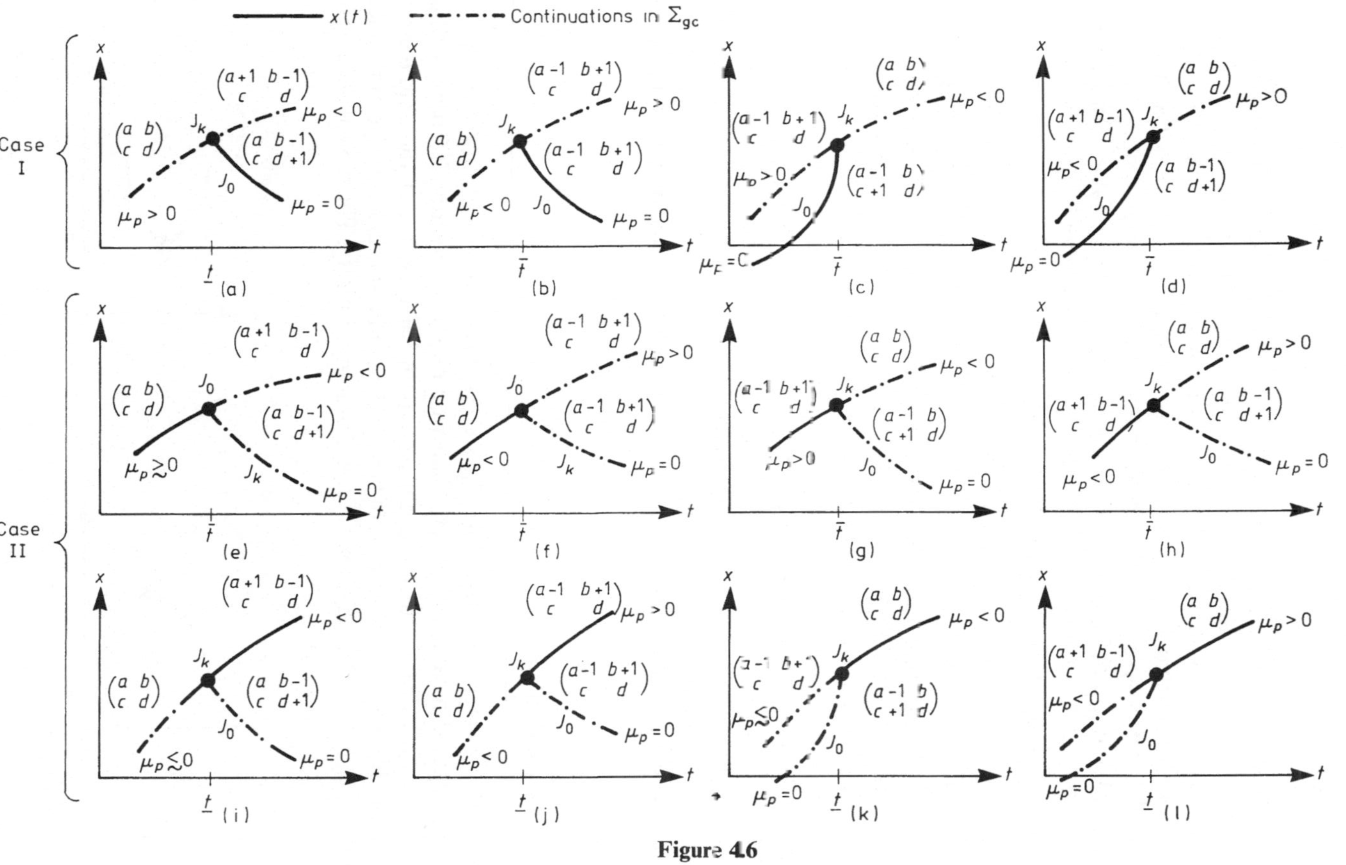

Figure 46

4.4.2 A point of type 3

Let $x(\bar{t}) = \bar{x}$ $(x(\underline{t}) = \underline{x})$ and $\bar{z} = (\bar{x}, \bar{t})$ $(\underline{z} = (\underline{x}, \underline{t}))$ be a point of type 3 and $[\underline{t}, \bar{t}]$ the considered local stability set. First, we note that the index set of active constraints is constant locally around $\bar{z}$. Then we have a look at the system (4.4.3) and note that the Jacobian with respect to (x, λ, μ, t) has the full rank around $\bar{z}$. Of course, for this we need only the weaker assumption that the functions f, h_i, g_j, $i \in I$, $j \in J$, are twice continuously differentiable. Furthermore, we observe that $\bar{z}$ is a quadratic turning point. This follows from the fact that $\bar{z}$ is a non-degenerate critical point for the problem

$$\max(\min)\{t \,|\, (x, t) \in \Sigma_{\mathrm{gc}}\}$$

with the linear function t. Therefore, the connected component C is quadratically approximable around $\bar{z}$ and parametrizable with respect to a parameter s, but not w.r.t. t (cf. e.g. Jongen *et al.* [116] and Section 4.2). The behaviour of C locally around $\bar{z}$ is illustrated in Figure 4.7.

Furthermore, we learn from Section 2.5 (type 3) that the linear indices $a = \mathrm{LI}$ and $b = \mathrm{LCI}$ remain constant around $\bar{z}$ whereas the quadratic index $c = \mathrm{QI}$ is changed exactly by one. If we restrict ourselves to Σ_{loc} as in Chapter 3, the curve of local minimizers stops at $\bar{z}$ and there is a continuation with a curve of stationary points with the index $c = 1$.

The above consideration implies that we can use the technique described in Section 4.2 in order to follow C numerically around $\bar{z}$.

4.4.3 A point of type 4

Let $x(\bar{t}) = \bar{x}$ $(x(\underline{t}) = \underline{x})$ and $\bar{z} = (\bar{x}, \bar{t})$ $(\underline{z} = (\underline{x}, \underline{t}))$ be a point of type 4 and $[\underline{t}, \bar{t}]$ the considered local stability set. Then we know from Section 2.5 (type 4) that $\bar{z}$ is also a quadratic turning point as a point of type 3, but the character is different. If we approach a point of type 4, then at least one Lagrange multiplier, say $v_p \in \{\lambda_i, i \in I, \mu_j, j \in J_0(\bar{z})\}$ tends to plus or minus infinity. The typical behaviour

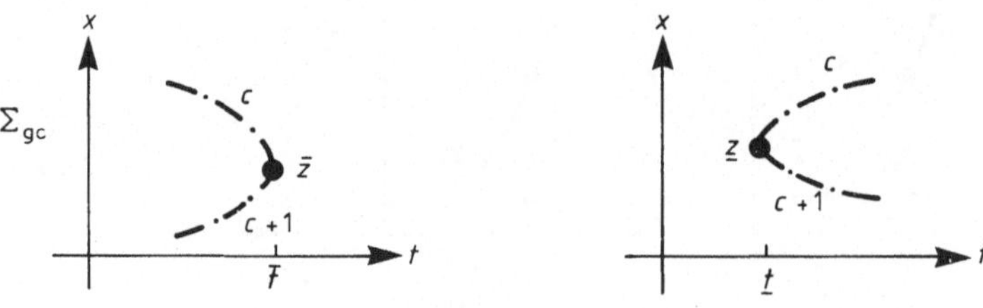

Figure 4.7

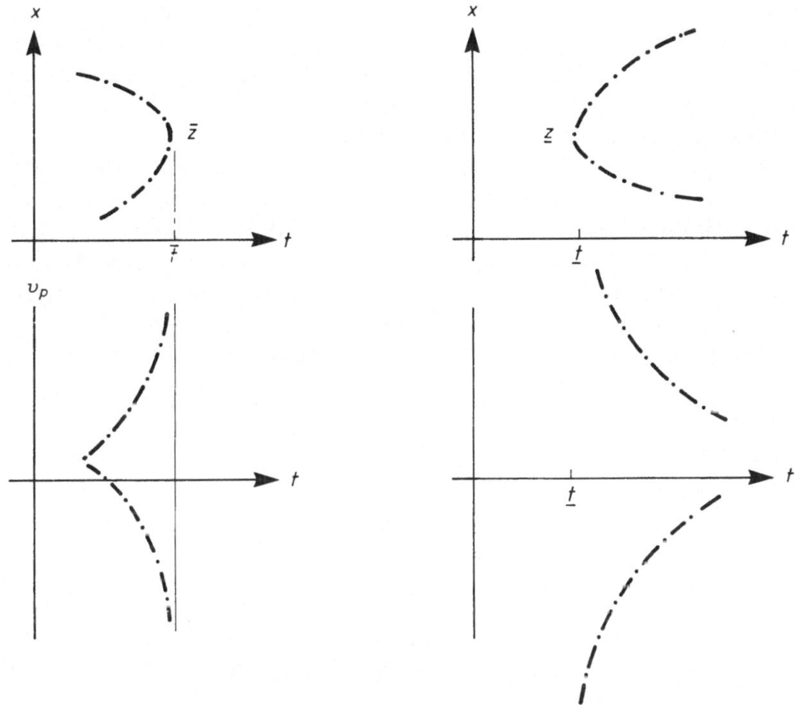

Figure 4.8

of C locally around $\bar{z}$ and the behaviour of the Lagrange multiplier v_p is illustrated in Figure 4.8 (for $\underline{t}$ and $\bar{t}$).

We see that the behaviour of the Lagrange multipliers indicates the approach to a point of type 4 in distinction to the approach to a point of type 3.

Concerning the change of the character of the critical points around $\bar{z}$, we refer to the index relations in Figure 2.15, e.g. a local minimizer turns into a local maximizer.

From the behaviour of $\lim_{t\to\bar{t}_-} v_p(t)$ it follows that we cannot use the system (4.4.3) in order to follow C numerically around $\bar{z}$. For this reason we propose (cf. Poore and Tiahrt [176]) the consideration of the following system for

$$t\in[\bar{t}-\delta_1,\bar{t}]\cup[\bar{t}-\delta_2,\bar{t}]:\ \lambda D_x f(x,t)+\sum_{i\in I}\lambda_i D_x h_i(x,t)+\sum_{j\in J_0}\mu_j D_x g_j(x,t)=0,$$

$$h_i(x,t)=0,\qquad i\in I,$$
$$g_j(x,t)=0,\qquad j\in J_0,$$
$$\lambda^2+\sum_{i\in I}\lambda_i^2+\sum_{j\in J_0}\mu_j^2=1,\qquad\qquad(4.4.4)$$

instead of the system (4.4.3). The choice of δ_1 with $0 < \delta_1$ depends on the behaviour of $v_p(t)$ in $(\underline{t}, \bar{t})$ for increasing t and $\delta_2 > 0$ was chosen sufficiently small. The transition from system (4.4.3) to (4.4.4) can be interpreted as a desingularization and a compactification, respectively, by means of a homogenization. Here, the idea of the notion of g.c. points is more obvious, because $\lambda(t) \to \underline{t} \to \bar{t}_- 0$. It is easy to see that the Jacobian with respect to (x, λ, μ, t) of the system (4.4.4) around $\bar{z}$ has the full rank (cf. e.g. Jongen *et al.* [116]) for $\lambda \neq 0$. Then we can use the technique described in Section 4.2 in order to follow C numerically around $\bar{z}$ using the following arguments: If we compare (4.4.3) and (4.4.4), we will see that the multipliers in (4.4.4) are bounded. Furthermore, for $t \in (\underline{t}, \bar{t})$ and $t \in (\bar{t} - \delta_2, \bar{t})$ we have the following properties:

To each $(\bar{x}(t), \bar{\lambda}(t), \bar{\lambda}_i(t), i \in I, \bar{\mu}(t), t)$ as a solution of (4.4.4) there correspond uniquely determined numbers $\hat{\lambda}_i(t), i \in I, \hat{\mu}_j(t), j \in J_0$, satisfying together with $(\bar{x}(t), t)$ the system (4.4.3) (by setting $\hat{\lambda}_i(t) := \bar{\lambda}_i(t)/\bar{\lambda}(t), i \in I, \hat{\mu}_j(t) := \bar{\mu}_j(t)/\bar{\lambda}(t), j \in J_0$), and, vice versa, to each $(\bar{x}(t), \hat{\lambda}_i(t), i \in I, \hat{\mu}(t), t)$ as a solution of (4.4.3) there correspond uniquely determined numbers $\bar{\lambda}(t)$, $\bar{\lambda}_i(t)$, $i \in I$, $\bar{\mu}_j(t)$, $j \in J_0$, where $\bar{\lambda}(t) = 1/w$, $\bar{\lambda}_i(t) = \hat{\lambda}_i(t)/w$, $i \in I$, $\bar{\mu}_j(t) = \hat{\mu}_j(t)/w$, $j \in J_0$, with

$$w = \left(1 + \sum_{i \in I} \hat{\lambda}_i(t) + \sum_{j \in J_0} \hat{\mu}_j \right)^{1/2},$$

satisfying together with $(\bar{x}(t), t)$ the system (4.4.4). For $t = \bar{t}$, the system (4.4.3) has no solution, but by using the properties of type 4 (cf. Section 2.5) the system (4.4.4) has a solution of the following form: $(\bar{x}, \bar{t})$ with the multipliers $\bar{\lambda}(\bar{t}) = 0$ and $\bar{\lambda}_i(\bar{t}), i \in I, \bar{\mu}_j(\bar{t}), j \in J_0$ (not all vanishing and unique up to a common multiple).

Of course, like for a point of type 3, we need only the weaker assumption that the functions $f, h_i, i \in I, g_j, j \in J$, are twice continuously differentiable for these considerations.

4.4.4   A point of type 5

Let $x(\bar{t}) = \bar{x}$ and $\bar{z} = (\bar{x}, \bar{t})$ be a point of type 5 and $[\underline{t}, \bar{t}]$ the considered local stability set. Such a point is mainly characterized by

$$|I| + |J_0(\bar{z})| = n + 1. \tag{4.4.5}$$

Since $m = |I| < n$, (4.4.5) implies that $|J_0(\bar{z})| \geqslant 2$. After renumbering we assume that $J_0(\bar{z}) = \{1, \ldots, p\}$. Then we know from Section 2.5 that the set Σ_{gc}, locally around $\bar{z}$, is equal to the union

$$\bigcup_{q=1}^{p} M_q^*,$$

where

$$M_q^* := \{(x,t) \in M_q \mid g_q(x,t) \leqslant 0\}$$

and

$$M_q := \{(x,t) \in \mathbb{R}^n \times \mathbb{R} \mid h_i(x,t)=0, i \in I, g_j(x,t)=0, j \in J_0(\bar z) - \{q\}\}, \qquad q \in \{1,\ldots,p\}.$$

Furthermore, locally around $\bar z$, the set M_q is a one-dimensional C^3 manifold, $q \in \{1,\ldots,p\}$, and $\bar z$ is a non-degenerate critical point for

$$P^{J_q}(t): \quad \min\{f(x,t) \mid h_i(x,t)=0, i \in I, g_j(x,t)=0, j \in J_0(\bar z)-\{q\}\},$$

$$t \in \mathbb{R}, \text{ for all } q \in J_0(\bar z).$$

Then, there exists a function $x^{J_q} \in C^3((\bar t - \varepsilon_q, \bar t + \varepsilon_q), \mathbb{R}^n)$, $\varepsilon_q > 0$, for each $q \in J_0(\bar z)$, and $(x^{J_q}(t), t)$ is a non-degenerate critical point for $P^{J_q}(t)$, $t \in (\underline t - \varepsilon, \bar t + \varepsilon)$ and $x^{J_q}(\bar t) = \bar x$.

Then we get all possible continuations at $\bar z$ by testing the feasibility of $x^{J_q}(t)$ for $t \in [\bar t - \varepsilon_q, \bar t + \varepsilon_q]$.

4.5 THE ALGORITHM PATH III

Algorithm PATH III

We start the algorithm with the point $(x^0, t_0) \in \Sigma_{gc}^1$, and want to describe numerically the connected component $C(x^0, t_0) \mid [t_A, t_B], t_A < t_0 < t_B$.

1. $(x^{J_0}(\bar t), \lambda^{J_0}(\bar t), \mu^{J_0}(\bar t))$ is computed by applying PATH II to the system (4.4.3). If $t_i = t_0$ and $\| x(t^i) - x^0 \| < \varepsilon$, ε sufficiently small, then STOP.

2. If we have case I of Theorem 4.4.1, then $J_0 := J_0 \cup \{j_0\}$; goto 8.

3. If we have case IIa of Theorem 4.4.1, then $J_0 := J_0$; goto 9.

4. If we have case IIb of Theorem 4.4.1, then $J_0 := J_0 \setminus \{p\}$; goto 10.

5. If at least one $v_p \in \{\lambda_i, i \in I, \mu_j, j \in J_0(\bar z)\}$ tends to $+\infty$ or $-\infty$, then switch to system (4.4.4) applying PATH II and, after passing the point of type 4, goto 9.

6. If $|I| + |J_0(\bar z)| = n + 1$, then goto 7, else goto 1.

7. For all $q \in J_0(\bar z)$ do $J_q := J_0 \setminus \{q\}$, apply PATH II to (4.4.3) with $J_0 = J_q$ and t in both orientations.
   If $x^{J_0}(t)$ is a feasible point for $\bar t < t < \bar t + \varepsilon$ or $\bar t - \varepsilon < t < \bar t$, then $J_0 := J_q$ and goto 9.

8. Solve (4.4.3) by PATH II with both orientations of t. Let $\tilde t$ be the next point of the discretization. Do $\bar t := \tilde t$ and goto 1.

9. Solve (4.4.3) by PATH II with the same orientation of t. Let $\tilde{t}$ be the next point of the discretization. Do $\bar{t} := \tilde{t}$ and goto 1.

10. Solve (4.4.3) by PATH II with either the same or the opposite orientation of t. Let $\tilde{t}$ be the next point of the discretization. Do $\bar{t} := \tilde{t}$ and goto 1.

REMARK 4.5.1   From the considerations in Section 4.4, it follows that the algorithm PATH III generates a numerical approximation of the connected component $C|[t_A, t_B]$ in the set Σ_{gc}. This version of the algorithm PATH III is not implementable in the strongest sense if points of type 5 occur (we need the active index set at such a point exactly). However, a stronger version is possible.

The following examples illustrate this algorithm.

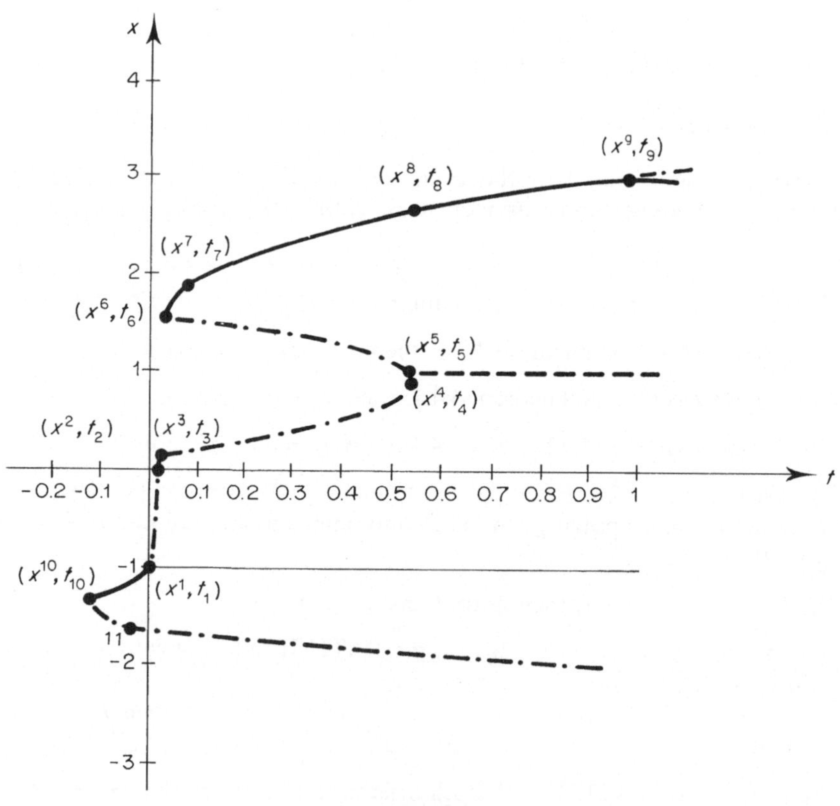

Figure 4.9

Example 4.5.1

$P(t)$: $\min\{f(x,t)\,|\,g_j(x,t)\leqslant 0, j=1,2,3,4,5\}$, $t\in[-1,1]$

where

$$f(x,t):=\tfrac{1}{5}x^5-\tfrac{1}{4}x^4-\tfrac{7}{3}x^3+\tfrac{1}{2}x^2+6x,$$
$$g_1(x,t):=\tfrac{1}{5}x^5-\tfrac{1}{4}x^4+0.6-30t,$$
$$g_2(x,t):=\tfrac{1}{2}x^2-1.5-4t,$$
$$g_3(x,t):=-\tfrac{7}{3}x^3+6x-0.5-6t,$$
$$g_4(x,t):=-x-2.2,$$
$$g_5(x,t):=x-5.$$

The application of the algorithm PATH III leads to the following geometrical interpretation (cf. Fig. 4.9 and Table 4.3)

Table 4.3   Singularities

i	t_i	x^i	TYPE	
1	0.005	-1	2	
2	0.02	0	4	
3	0.019 999 9	0.103 745 9	5	(MFCQ is satisfied)
4	0.533 88	0.925 82	4	
5	0.527 780 7	1	2	
6	0.031 070 7	1.542 985 4	5	(MFCQ not satisfied)
7	0.076 643 7	1.900 826 5	5	(MFCQ is satisfied)
8	0.540 081 8	2.705 670 7	5	(MFCQ is satisfied)
9	0.965	3	2	
10	-0.125	$-1.360 036 7$	5	(MFCQ not satisfied)
11	-0.044	$-1.622 740 0$	5	(MFCQ is satisfied)

Example 4.5.2

$P(t)$: $\min\{f(x,t)\,|\,g_j(x,t)\leqslant 0, j=1,\ldots,5\}, t\in[-1,2]$

where

$$f(x,t):=-3.585\,207\,18\times10^{-3}x^9+0.014\,955\,357x^8+0.110\,830\,025x^7$$
$$-0.417\,708\,329x^6-1.049\,594\,88x^5+3.357\,812\,43x^4+3.391\,953\,17x^3$$
$$-7.955\,059\,4x^2-14.449\,602\,98x+3$$
$$g_1(x,t)=0.014\,955\,357x^8-1.049\,594\,88x^5+3.391\,953\,17x^3-7.955\,059\,4x^2$$
$$-1000+1000t$$

$$g_2(x,t) = -0.417\,708\,329x^6 - 3.585\,207\,18 \times 10^{-3}x^9 + 3.357\,812\,43x^4 - 30 - 20t$$

$$g_3(x,t) = 0.110\,830\,025x^7 - 14.449\,602\,98x + 3 - 9t$$

$$g_4(x) = x - 5$$

$$g_5(x) = -x - 5$$

Applying the algorithm PATH III we obtain Figure 4.10.

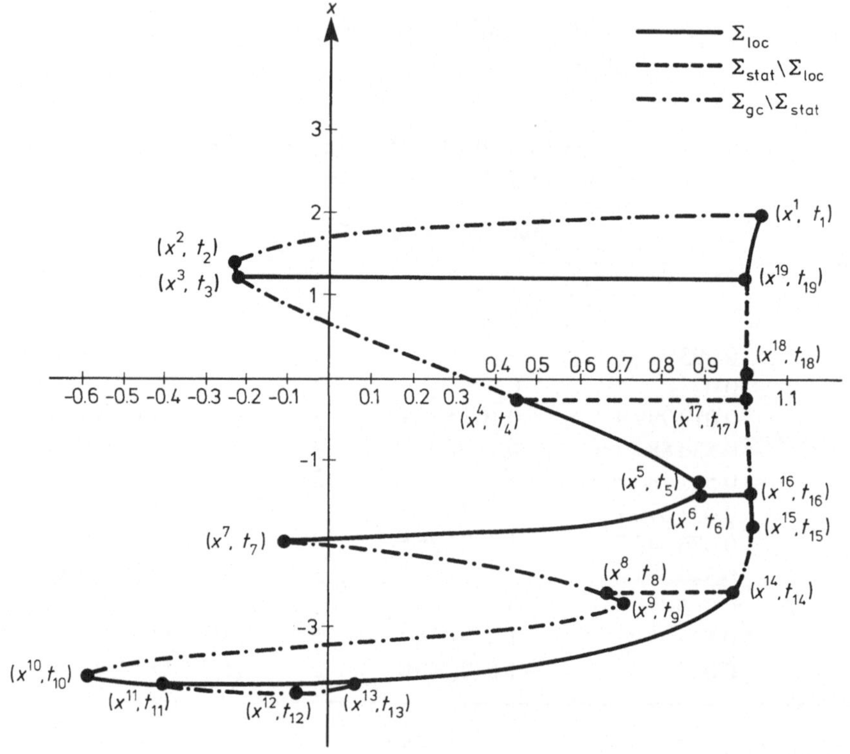

Figure 4.10

(x^i, t_i), $i = 3, 4, 6, 8, 11, 13, 14, 16, 17, 19$ are points of type 2, (x^i, t_i), $i = 5, 9, 10,$ 15, 18 are points of type 4, (x^i, t_i), $i = 1, 7$ are points of type 5 (where the MFCQ is violated) and (x^{12}, t_{12}) is a point of type 5 (where the MFCQ is satisfied).

5

Pathfollowing with Jumps in the Set of Local Minimizers and in the Set of Generalized Critical Points

5.1 PRELIMINARY OUTLINE

We consider the one-parametric optimization problem $P(t), t\in[0, 1]$, where $P(t)$ is taken as in Section 1.1, namely

$$P(t): \qquad\qquad \min\{f(x, t)\,|\,x\in M(t)\}, \qquad\qquad t\in[0, 1], \qquad\qquad (5.1.1)$$

where

$$M(t) = \{x\in\mathbb{R}^n\,|\,h_i(x, t) = 0, i\in I, g_j(x, t)\leqslant 0, j\in J\} \qquad\qquad (5.1.2)$$

with

$$I = \{1,\ldots,m\}, \qquad m < n, \qquad J = \{1,\ldots,s\}. \qquad\qquad (5.1.3)$$

Throughout this chapter we assume:

(A1) $(f, H, G)\in\mathscr{F}^{**}$,

(A2) $(x^0, 0)\in\Sigma^1_{\mathrm{gc}}$ is known or easy to compute,

(A3) for all $t\in[0, 1], M(t)$ is non-empty and there exists a compact set C containing $M(t)$.

As noted at several places in this book (cf. e.g. Chapter 1) we cannot expect to find a discretization of $[0, 1]$:

$$0 = t_0 < \cdots < t_i < t_{i+1} < \cdots < t_N = 1 \qquad\qquad (5.1.4)$$

and for each $t_i, i\in\{1,\ldots,N\}$, a local minimizer $(x(t_i), t_i), i = 1,\ldots,N$ (resp. stationary point or g.c. point) by using pathfollowing methods (described in Chapters 3 and 4) only (cf. e.g. Figures 1.1 and 5.1).

Example 5.1.1

$$\min \{ f(x, t) \mid x \in \mathbb{R} \}, \qquad t \in \mathbb{R},$$

where

$$f(x, t) = x \sin x - 4t \arctan [6t(x + 3.5)] + 4t \arctan [3(2t - 1)(x + 1)]. \quad (5.1.5)$$

Then the structure of Σ_{stat} is illustrated in Figure 5.1. The full curve stands for the curve of local minimizers and the dotted curve represents a curve of stationary points not being local minimizers.

Such a situation is typical of the non-convex case, in particular of applications in global optimization (cf. Section 6.3). For this reason we propose to jump from one connected component to another one. We try to compute a descent direction at a turning point by using the information on this special singularity; in particular we consider a smooth normal form of $M(t)$ and $f(x, t)$ (by a smooth local coordinate transformation). In Section 5.2 we discuss jumps in the set $\bar{\Sigma}_{\text{loc}}$. We note that we still do not have jumps for all cases. Therefore, we cannot find a discretization (5.1.4) and the corresponding local minimizers $x(t_i), i = 1, \ldots, N$, with certainty. This is not unexpected, since, when we have jumps in Σ_{loc} for

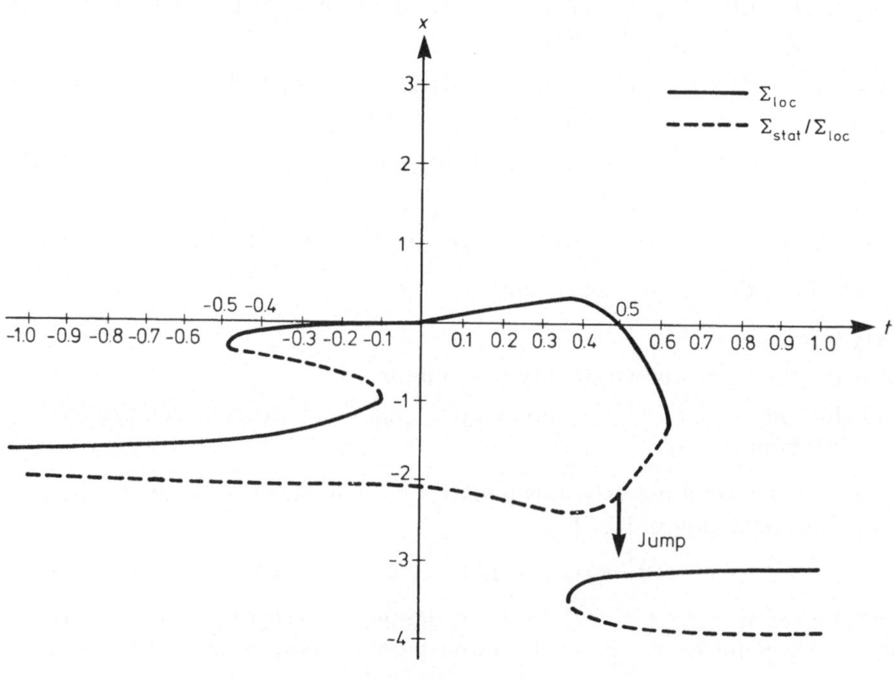

Figure 5.1

all singularities, the problem of global optimization is solved under the assumption (A1) (cf. Sections 1.2 and 6.3). This is the reason why we consider jumps from a connected component in Σ_{gc} to another one in Section 5.3. We find directions of descent from special points of $\Sigma_{gc}^1 \setminus \Sigma_{stat}$ and $\Sigma_{stat}^1 \setminus \Sigma_{loc}$ ($\Sigma_{stat}^1 := \Sigma_{gc}^1 \cap \Sigma_{stat}$) and in some cases also from points of $\bar{\Sigma}_{loc}$. So, we can try to find several (maybe all) connected components and their numerical tracing. We can use this approach if we are not successful with jumps in $\bar{\Sigma}_{loc}$ (Section 5.2). Of course, we have to realize the pathfollowing method described in Section 4.4 for the numerical representation of the connected component in Σ_{gc}. We note that we do not have a jump to another connected component in Σ_{gc} in all cases. Nevertheless, this approach could be useful for the applications in Chapter 6 (to find a discretization (5.1.4) and corresponding g.c. points $x(t_i)$), $i = 1, \ldots, N$, for the special embeddings). Furthermore, if the numerical description of all components in Σ_{gc} is of interest, this investigation could be useful, too. From this point of view it will be helpful to have an estimation of the number of connected components in Σ_{gc}, but this is still another open question.

5.2 JUMPS IN THE SET OF LOCAL MINIMIZERS AND THE ALGORITHM JUMP I

In this section we follow the articles by Jongen [112] and Guddat *et al.* [89, 90]. We assume that (A1) and (A3) are satisfied and specify (A2) in the following way:

(A2) $(x^0, 0) \in \Sigma_{loc}^1$ $(\bar{\Sigma}_{loc}^i := \bar{\Sigma}_{gc}^i \cap \bar{\Sigma}_{loc}, i = 1, \ldots, 5)$ is known.

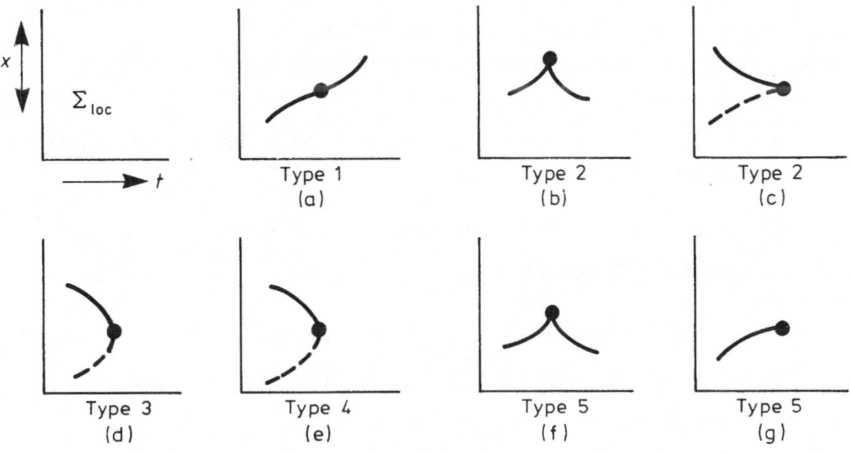

Figure 5.2

In order to simplify the proofs in this section we assume that the functions $f, h_i, g_j, i \in I, j \in J$, belong to $C^\infty(\mathbb{R}^n \times \mathbb{R}, \mathbb{R})$.

Under the assumption (A1) the local structure of $\bar{\Sigma}_{\mathrm{loc}}$ is described in Section 2.5 and illustrated in Figure 2.17. Since we will walk along branches of $\bar{\Sigma}_{\mathrm{loc}}$ for increasing values of t, the situation of Figures 2.17(d), (f), (h) and (k) will not occur. Therefore, only the local structure illustrated in Figure 5.2 is interesting for our consideration.

As in Figure 2.17, the point $\bar{z} = (\bar{x}, \bar{t})$ is identified by an exposed point, whereas the full curve stands for the curve of local minimizers and the dotted curve stands for the curve of generalized critical points not being local minimizers. We recall that the dotted curve in Figure 5.2(d)(e) represents even a curve of stationary points. We also have a curve of stationary points in Figure 5.2(e) if $J_0(\bar{z}) = \emptyset$.

5.2.1 Points of type 2

For the theoretical background we refer the reader to Sections 2.3 and 2.5, and for an example of a turning point of type 2 to Example 4.3.1. A g.c. point $\bar{z} = (\bar{x}, \bar{t})$ is of type 2 if the strict complementarity condition is violated for exactly one active inequality constraint, say having index $\tilde{j} \in J_0(\bar{z})$; however, LICQ is satisfied. The LICQ implies the existence of numbers $\bar{\lambda}_i, \bar{\mu}_j, i \in I, j \in J_0(\bar{z})$ (not all vanishing), such that

$$D_x f + \sum_{i \in I} \bar{\lambda}_i D_x h_i + \sum_{j \in J_0(\bar{z})} \bar{\mu}_j D_x g_j \big|_{\bar{z}} = 0. \tag{5.2.1}$$

The violation of the strict complementarity condition on $\tilde{j} \in J_0(\bar{z})$ means that the corresponding Lagrange multiplier $\bar{\mu}_{\tilde{j}}$ in (5.2.1) vanishes. Let $J_0(\bar{z}) = \{1, \ldots, p\}$. We refer to Section 2.5 for a detailed description.

First, we derive a normal form for the situation of type 2.

LEMMA 5.2.1 Let $\bar{z} = (\bar{x}, \bar{t})$ be a g.c. point of type 2. Then, there exists a C^∞ diffeomorphism $\psi: U \to V$ of the form

$$\psi: (x, t) \to (\psi_1(x, t), \psi_2(t)), \qquad \psi_2'(t) > 0, \tag{5.2.2}$$

sending $\bar{z}$ onto the origin such that:

(i) $$\psi[(M(t), t) \cap U] = (\mathbb{H}^p \times \mathbb{R}^q \times \{0_{n-p-q}\}, \psi_2(t)) \cap V \tag{5.2.3}$$

($\mathbb{H}^p$ is defined by (2.3.7) and $(M(t), t) := \{(x, t) \in \mathbb{R}^n \times \mathbb{R} \mid x \in M(t)\}$),

(ii) $$f \circ \psi^{-1}(y_1, \ldots, y_{p+q}, 0, \ldots, 0, v)$$

$$= \alpha \gamma(v)(y_1 - \beta v)^2 + \delta(v) + \sum_{i=2}^{p} (\pm y_i) + \sum_{j=1}^{p+q} (\pm y_j^2) \tag{5.2.4}$$

with $\alpha, \beta \in \{-1, +1\}$, and $\gamma(\cdot), \delta(\cdot)$ smooth functions around the origin, and $\gamma(0) > 0$.

Proof By means of the splitting argument performed in [121, pp. 525–9] we may already assume that, in new smooth coordinates according to (5.2.2), we have the following situation:

$$f(x, t) = F(x_1, t) + \sum_{i=2}^{p} (\pm x_i) + \sum_{j=p+1}^{p+q} (\pm y_j^2), \tag{5.2.5}$$

$$x_i \geqslant 0, \qquad i = 1, \ldots, p, \tag{5.2.6}$$

where the origin is a point of type 2. So, in our analysis we may focus on the situation of *one variable* x, i.e. we look at $F(x, t), x \geqslant 0$, in a neighbourhood of the origin in $\mathbb{R} \times \mathbb{R}$. In the next local change of coordinates the feasible set $\{x \in \mathbb{R} \mid x \geqslant 0\}$ is transformed to itself, and so we take coordinate transformations of the following form (cf. (5.2.2)):

$$\psi(x, t) = (x, \psi_1(x, t), \psi_2(t)), \tag{5.2.7}$$

where

$$\psi_1(0) > 0 \qquad \text{and} \qquad \psi_2'(0) > 0. \tag{5.2.8}$$

Note that, in view of the special structure of ψ in (5.2.7), the origin $x = 0$ is mapped onto itself. Since we consider a point of type 2, its characterization (cf. [9]) implies

$$F_x(0) = 0 \qquad F_{xx}(0) \neq 0, \qquad F_{xt}(0) \neq 0, \tag{5.2.9}$$

where F_x, F_{xx}, F_{xt} denote the corresponding partial derivatives.

By virtue of (5.2.9) the application of the implicit function theorem yields the existence of a local smooth function $t \to \eta(t)$ such that

$$F_x(\eta(t), t) \equiv 0, \qquad \eta(0) = 0. \tag{5.2.10}$$

Put

$$\phi(u) = F(\eta(t) + u[x - \eta(t)], t). \tag{5.2.11}$$

Then, we have

$$\phi(0) = F(\eta(t), t), \qquad \phi(1) = F(x, t) \tag{5.2.12}$$

and it follows that

$$\phi(1) - \phi(0) = \int_0^1 \frac{d\phi(u)}{du} du = \int_0^1 F_x(\eta(t) + u[x - \eta(t)], t)[x - \eta(t)] du$$

$$= [x - \eta(t)]G(x, t). \tag{5.2.13}$$

From (5.2.10), (5.2.12) and (5.2.13) we see that $G(\eta(t), t) \equiv 0$, and hence, putting

$$\xi(u) = G(\eta(t) + u[x - \eta(t)], t), \tag{5.2.14}$$

we obtain, similarly as in (5.2.13),

$$G(x,t) = \xi(1) = \xi(1) - \xi(0) = \cdots = [x - \eta(t)]H(x,t). \qquad (5.2.15)$$

Finally, a combination of (5.2.12), (5.2.13) and (5.2.14) yields:

$$F(x,t) = F(\eta(t),t) + [x - \eta(t)]^2 H(x,t). \qquad (5.2.16)$$

Using the chain rule for differentiating $t \to F_x(\eta(t),t)$, it follows from (5.2.9) that

$$\beta := \mathrm{sign}\left(\frac{\mathrm{d}}{\mathrm{d}t}\eta(0)\right) \neq 0.$$

Taking $t \to \beta\eta(t)$ as a new t coordinate, we may assume that F has the following form $(\delta(t) := F(\eta(t),t))$:

$$F(x,t) = \delta(t) + (x - \beta t)^2 w(x,t). \qquad (5.2.17)$$

Finally, define

$$\begin{aligned} &y = (x - \beta t)(|w(x,t)|)^{1/2}/(|w(0,t)|)^{1/2} + \beta t \\ &v = t. \end{aligned} \qquad (5.2.18)$$

Putting $\alpha = \mathrm{sign}\, w(0)$ and $\gamma(v) = |w(0,v)|$, it is easily checked that, at last, formula (5.2.18) fits our aim. $\qquad \square$

Recall that we are interested in walking along a path of local minimizers for $P(t)$ as the parameter t increases. From (5.2.3) and (5.2.4) it follows that the normal form (5.2.4) has only *positive* linear/quadratic terms for $i = 2, \ldots, p$ and

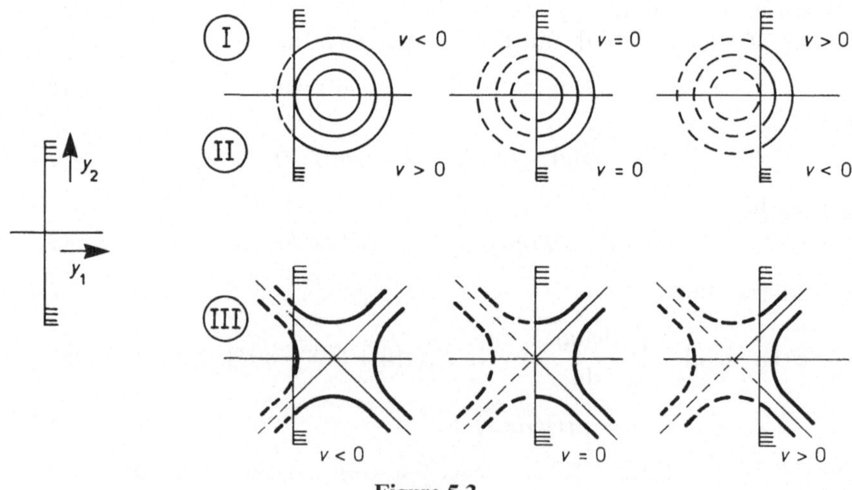

Figure 5.3

$j = p + 1, \ldots, p + q$, so we have

$$f \circ \psi^{-1}(y_1, \ldots, y_{p+q}, 0, \ldots, 0, v) = \alpha\gamma(v)(y_1 - \beta v)^2 + \delta(v) + \sum_{i=2}^{p} y_i + \sum_{j=p+1}^{p+q} y_j^2.$$

$$(5.2.19)$$

Now, it is easily seen that there are precisely three possibilities for α, β:

$$
\begin{array}{lll}
\text{I} & \beta = -1, \alpha = +1, & \\
\text{II} & \beta = +1, \alpha = +1, & (5.2.20) \\
\text{III} & \beta = -1, \alpha = -1. &
\end{array}
$$

In Figure 5.3 the above possibilities are depicted in the case $p = q = 1$, $\gamma(v) \equiv |v|$; according to (5.2.14), the y dimension equal 2, and the inequality constraint is $y_1 \geq 0$. Some level lines in the y space are depicted for negative, zero and positive values of v.

In Figure 5.4 the above possibilities I, II and III are depicted from the point of view of Figures 2.12 and 2.13.

In terms of the original functions f, h_i, g_j, the difference between the situations I, II (cf. (5.2.20)) on the one hand and III on the other can be explained as follows. Let $\tilde{j} \in J_0(\bar{z})$ again denote the index of the active inequality constraint for which the Lagrange multiplier $\mu_{\tilde{j}}$ in (5.2.1) vanishes. Let the Lagrangian L and the tangent space $T(\bar{z})$ be defined according to (2.5.12) and (2.5.10), respectively, and consider the *larger tangent space* $\tilde{T}(\bar{z})$ defined by (2.5.11).

Then, in the situations I, II both $D^2L(\bar{z})/T(\bar{z})$ and $D^2L(\bar{z})/\tilde{T}(\bar{z})$ are positive definite; however, in III we have $D^2L(\bar{z})/T(\bar{z})$ is positive definite, but $D^2L(\bar{z})/\tilde{T}(\bar{z})$ is not positive definite. In situation III it holds that

$$\min \{\xi^T D^2 L(\bar{z})\xi \mid \|\xi\| = 1, \xi \in \tilde{T}(\bar{z}), Dg_{\tilde{j}}(z)\xi \geq 0\} < 0. \qquad (5.2.21)$$

So, in the situations I, II, we can continue following a branch of local minimizers either by treating the constraint $g_{\tilde{j}}$ as an equality constraint (I) or

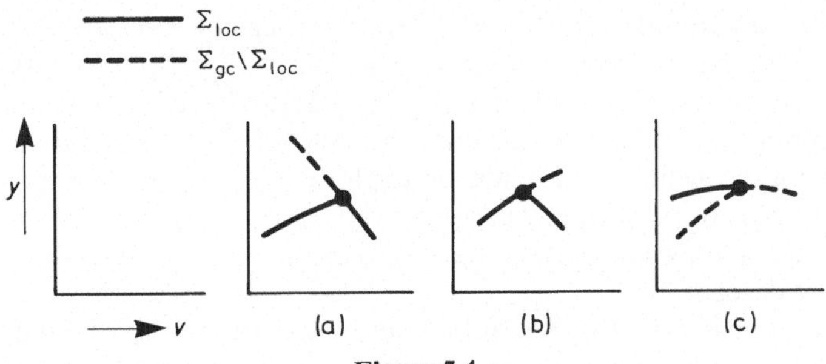

Figure 5.4

by omitting g_j as a constraint (II) (cf. the algorithms PATH I and PATH II in Chapters 3 and 4).

In situation III the branch of local minimizers stops at $\bar{z}$ as t increases. However, it is easily seen that, for $z=(x,t)\in\Sigma_{gc}$ in a sufficiently small neighbourhood of $\bar{z}$, the point x cannot be a global minimizers for $P(t)$. In view of (5.2.21), at the point $\bar{z}$ we can find a (quadratic) descent direction ξ for the problem $P(t)$, and we can find a local minimizers $\tilde{x}$ for $P(\bar{t})$. Of course, $\tilde{x}\neq\bar{x}$. Then, we can continue tracing the branch of local minimizers through the point $(\tilde{x},\bar{t})$, and so a jump to another branch has been realized.

REMARK 5.2.1   We also refer to a proposal of Rückmann for a special direction of descent (cf. [199] and [89]). This vector can be computed in the same way as in formula (5.2.22) for a point of type 3. In order to find a vector ξ with the property (5.2.21), we have to compute a global minimizer (in the extreme case) for a non-convex quadratic optimization problem ($\|\cdot\|$ is chosen in a suitable way) (cf. e.g. [167] for a solution algorithm), but this could be difficult. The proposal of Rückmann is a good alternative.

5.2.2   The jump at a point of type 3

A g.c. point $\bar{z}=(\bar{x},\bar{t})$ is of type 3 if the LICQ and strict complementarity holds, but one eigenvalue of $D^2L(\bar{z})/T(\bar{z})$ vanishes. We refer to Section 2.5 for a detailed description and to Example 5.1.1.

In the (x,t) space the set Σ_{gc} exhibits a quadratic turning point (with respect to the t direction) (for details cf. Section 4.4).

So, we can still compute along the turning point by means of a pathfollowing method (cf. PATH II in Chapter 4). However, upon passing the turning point a local minimizer turns into a point belonging to $\Sigma_{stat}^1\setminus\Sigma_{loc}$ since (exactly) one eigenvalue of $D^2L(\bar{z})|T(\bar{z})$ changes from a positive to a negative sign; see also [121, Chapter 10] for a detailed discussion. From this we conclude that the point x is no local minimizer for $P(\bar{t})$. Since the feasible set $M(t)$ was assumed to be compact, the existence of a global minimizer $\tilde{x}$ for $P(\bar{t})$ is guaranteed. So, if we use a descent method starting at $\bar{x}$, we can arrive at a local minimizer $\hat{x}$, and the point $(\hat{x},\bar{t})$ lies on another branch of $\bar{\Sigma}_{loc}$. Starting at $(\hat{x},\bar{t})$, a pathfollowing procedure can again be exploited. The actual jump consists of the transition from $(\bar{x},\bar{t})$ to $(\hat{x},\bar{t})$.

The problem now consists of finding—in an effective way—a tangential direction of descent.

Our proposal is based on the following observation: regard Figure 5.5. Let t be near $\bar{t}, t<\bar{t}$, and let $x_m(t)$ be the local minimizer and $(x_s(t),t)\in\Sigma_{stat}^1\setminus\Sigma_{loc}$.

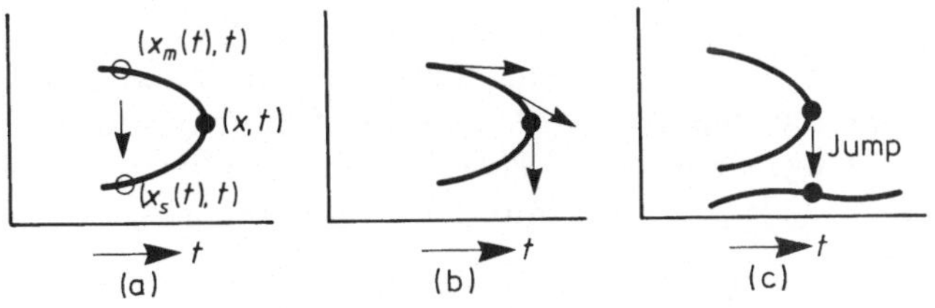

Figure 5.5

Then, as t tends to $\bar{t}$, the vector

$$u(t) := [x_s(t) - x_m(t)] / \| x_s(t) - x_m(t) \| \tag{5.2.22}$$

tends to a tangential vector, say $\bar{u}$, which is a direction of (higher-order) descent. Hence, for t near $t, t < \bar{t}$, the vector $x_s(t) - x_m(t)$ provides an approximately tangential direction of descent.

This becomes obvious as follows. In view of the validity of the LICQ and the strict complementarity, we can restrict ourselves (in new local C^3 coordinates, cf. Section 2.3) to the unconstrained case. Using the characteristics of a point of type 3 (cf. Section 2.5), we then deal with a C^3 function $f(x, t)$ having the following properties at $\bar{z} = (\bar{x}, \bar{t})$:

(i) $D_x f(\bar{x}, \bar{t}) = 0$,

(ii) $D_x^2 f(\bar{x}, \bar{t})$ is positive semidefinite with exactly one vanishing eigenvalue.

Let $D_x^2 f(\bar{x}, \bar{t})\xi = 0, \| \xi \| = 1$. Then we have

(iii) $D_x^3 f(\xi, \xi, \xi) \neq 0$ and $D_t(D_x f \cdot \xi) \neq 0$.

From a Taylor expansion for $f(\cdot, \bar{t})$ at $\bar{x}$ we see that either ξ or $-\xi$ is a direction of cubic descent. Note that v is ξ or $-\xi$. A further clarification can be given by applying singularity theory. In fact, we assume that f is a C^∞ function (the function f can be locally approximated in the C^3 sense arbitrarily well by means of a C^∞ function $\tilde{f}$ having the same partial derivatives up to order 3 at the point $(\bar{x}, \bar{t})$). Then the preceding conditions (i)–(iii) and unfolding theory (cf. [121]) provide the existence of a local C^∞ coordinate change

$$(y, v) = \phi(x, t), \tag{5.2.23}$$

having the structure

$$\phi(x, t) = (\psi(x, t), \eta(t)), \qquad \eta'(\bar{t}) > 0, \tag{5.2.24}$$

such that $\phi(\bar{x}, \bar{t}) = 0$ and

$$f \circ \phi^{-1}(y, v) = y_1^3 + vy_1 + \sum_{i=2}^{n} y_i^2 + \delta(v), \tag{5.2.25}$$

where $\delta(v)$ represents the functional value at $y = 0$. We note once more that $f \circ \phi^{-1}(y, v)$ is nothing else but the function f in the (fibrewise) new coordinates. From (5.2.25) we see that $D_y f \circ \phi^{-1}$ vanishes iff both $y_i = 0, i = 2, \ldots, n$, and $3y_1^2 + v = 0$ (defining a parabola). Then we obtain $y_1^{1,2} = \pm(-v/3)^{1/2}, v < 0$. Substituting $y^{1,2} = (y_1^{1,2}, 0, \ldots, 0)^T$ into (5.2.25) we obtain

$$f \circ \phi^{-1}(y^1, v) = \tfrac{2}{3}v(-v/3)^{1/2} + \delta(v),$$
$$f \circ \phi^{-1}(y^2, v) = -\tfrac{2}{3}v(-v/3)^{1/2} + \delta(v).$$

Furthermore, we see that the Hessian $H(y, v)$ with respect to y of $f \circ \phi^{-1}(y, v)$ is

$$H(y^1, v) = \begin{pmatrix} 6y_1^1 & 0 & \cdots & 0 \\ 0 & 2 & \cdots & 0 \\ \vdots & \vdots & \ddots & \\ 0 & 0 & & 2 \end{pmatrix}$$

and $H(y^1, v)$ is positive definite (since $v < 0$), whereas $H(y^2, v)$ is indefinite. Then, y^1 is a local minimizer and y^2 is a saddle point. Now we consider

$$\bar{u}(v) := \frac{y^2 - y^1}{\| y^2 - y^1 \|}$$

(compare (5.2.22)) and obtain $\bar{u}(v) = (-1, 0, \ldots, 0)$. Furthermore, we see that $H(\bar{y})\bar{u}(v) = 0$ and

$$\frac{\partial^3 f(\bar{u}(v))}{\partial y_i \partial y_j \partial y_k} = \begin{cases} -6 & \text{if } i = j = k = 1, \\ 0 & \text{otherwise.} \end{cases}$$

Using Taylor's expansion yields that $\bar{u}(0)$ is a direction of cubic descent. For u varying from negative to positive values we have depicted the level lines of $f \circ \phi^{-1}(\cdot, v)$ in Figure 5.6. This clarifies the direction $\bar{u}$ in the preceding discussion.

Since we suppose the feasible set $M(\bar{t})$ for $P(\bar{t})$ to be compact, we can jump from $\bar{z}$ to another branch of local minimizers (note that for $(x, t) \in \Sigma_{gc}$, in a neighbourhood of $\bar{z}$ there cannot be a global minimizer for $P(t)$). In fact, at $\bar{z}$ there exists a unique (tangential) direction of (cubic) descent; in (5.2.25) this will be the negative of the $(p+1)$th unit vector. Descending in that direction within

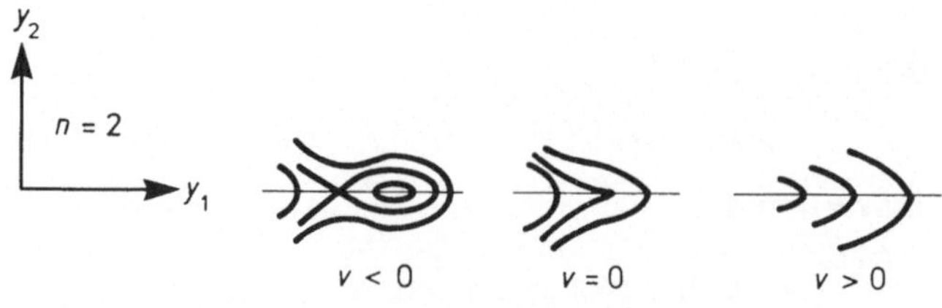

Figure 5.6

$M(\bar{t})$ leads to a local minimizer $\tilde{x}$ ($\tilde{x} \neq \bar{x}$), and we can continue tracing the branch of a local minimizer through the point $(\tilde{x}, \bar{t}) \in \Sigma_{gc}$. In the original coordinates the latter direction of descent can be—approximately—obtained as follows, cf. Figure 5.5(a). For $t < \bar{t}$ the intersection of Σ_{gc} with the hyperplane '$t = $ const.' consists, in a neighbourhood of $\bar{z}$, of exactly two points, one of them being a local minimizer $(x_m(t), t)$ and the other one being a saddle point $(x_s(t), t)$. Then, the vector $[x_s(t) - x_m(t)]/\|x_s(t) - x_m(t)\|$ converges to the desired direction of descent as t tends to $\bar{t}$. The latter vector is also the limit of the unit 'forward' tangential vector along Σ_{gc} taken at the points $(x_m(t), t)$, as t tends to $\bar{t}$ (cf. Figure 5.5(b)).

5.2.3 Points of type 4 and type 5

A g.c. point $\bar{z} = (\bar{x}, \bar{t})$ is of type 4 or type 5 if the LICQ is violated at $\bar{z}$. For a detailed description we refer again to Section 2.5. Example 5.2.1 illustrates a point of type 4, and Example 5.2.2 a point of type 5, where the MFCQ is violated.

Example 5.2.1

Let $h(x) = x^3 - 3x^2 - x + 7$ and $x^0 = 4$, and let us consider the problem

$$\min\{(x - x^0)^2 \,|\, h(x) + (t - 1)h(x^0) = 0\}.$$

The behaviour of the g.c. points and the corresponding Lagrange parameter λ is depicted in Figure 5.7, where $z^0 = (4, 0), z^1 = (2.155, 0.951\,537)$, $z^2 = (-0.154\,7005, 0.627\,4105)$, $z^3 = (-1.382\,9758, 1)$.

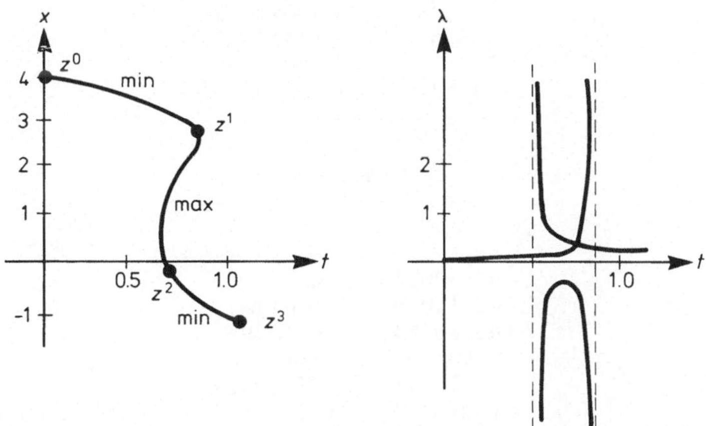

Figure 5.7

The curve connecting z^0 and z^1 corresponds to Table 5.1 for the Lagrange multiplier λ (cf. (5.2.1)).

The curve connecting z^1 and z^2 corresponds to Table 5.2 for the Lagrange parameter λ.

The curve connecting z^2 and z^3 corresponds to Table 5.3 for the Lagrange parameter λ.

Table 5.1

x	λ
2.155	$+\infty$
2.108 902 5	18.496 12
2.208 550 9	4.692 285 9
2.295 336 1	2.688 785 6
2.836 823 1	0.190 007
3.087 873 8	0.100 480 9
3.991 988 7	0.034 775 8
4.0	0.0

Table 5.2

x	λ
$-0.154\,700\,5$	$-\infty$
0.0	-4.0
0.217 902 2	$-1.746\,952\,3$
0.552 613 4	$-1.014\,075\,6$
0.810 807	$-0.819\,2925$
1.050 031 3	$-0.738\,879\,2$
1.571 718 2	$-0.804\,222\,6$
2.061 628	$-3.132\,276\,9$
2.1	$-8.135\,135$
2.155	$-\infty$

Table 5.3

x	λ
$-0.154\,700\,5$	$+\infty$
$-0.211\,107\,9$	10.518 647
$-0.296\,855\,4$	6.396 397 3
$-0.490\,77$	1.683 711 2
$-0.738\,418\,8$	0.935 281 9
$-1.029\,536\,3$	0.533 381 1
$-1.309\,901$	0.442 245

Furthermore, we note that the points of the curves connecting z^0 and z^1 (resp. z^2 and z^3) are local minimizers and the points of the curve connecting z^1 and z^2 are local maximizers. This example is typical for applications in global optimization (cf. (1.2.2) and Section 6.3) and is constructed by using the embedding (1.2.7) and (1.2.8).

Example 5.2.2

$$\min\{f(x)\,|\,g_i(x,t)\leqslant 0, i=1,2,3\}, \qquad t\in\mathbb{R},$$

where

$$f(x) = -x_1, \qquad g_1(x) = -x_1, \qquad g_2(x) = -x_2,$$
$$g_3(x) = x_1 + x_2 + t - 1.$$

We see that $(x_1(t), x_2(t)) = (1-t, 0)$ is a global minimizer for all $t\in[0,1]$, and the corresponding Lagrange parameters $\mu_1(t), \mu_2(t)$ and $\mu_3(t)$ are the following ones:

$$\mu_1(t) = 0, \quad \text{if} \quad t\in[0,1) \qquad \mu_1(1)\in[0, +\infty),$$
$$\mu_2(t) = 1, \quad \text{if} \quad t\in[0,1) \qquad \mu_2(1)\in[1, +\infty),$$
$$\mu_3(t) = 1, \quad \text{if} \quad t\in[0,1) \qquad \mu_3(1)\in[1, +\infty).$$

This behaviour is depicted in Figure 5.8.

We note that the feasible set $M(t)$ cannot be smoothly transformed into the standard set as in (5.2.3).

Thus, we have to derive a new local normal form for $M(t)$ in which the parameter t plays an essential role. Such normal forms are studied in [115] and [121, Chapter 10]. It turns out that, for increasing $t, M(t)$ behaves locally like (lower-, upper-) level sets of an objective function in the neighbourhood of a non-degenerate critical point (cf. Section 2.3). In fact, there exists a smooth coordinate transformation ψ satisfying (5.22) such that $M(t)$ locally transforms

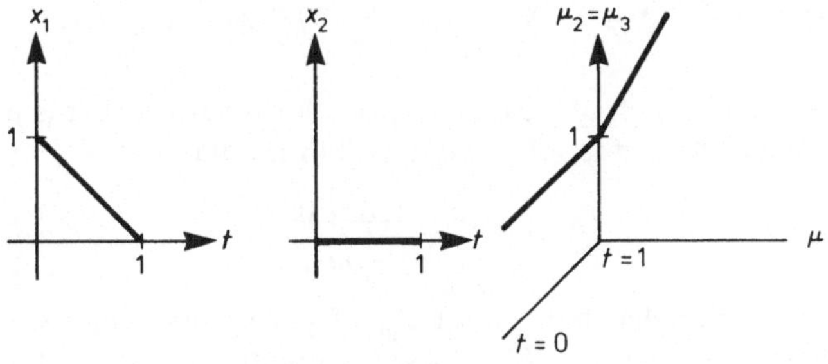

Figure 5.8

as follows (we put $(y,v) = \psi(x,t)$ and write down the (in)equalities defining $M(t)$ locally in the new coordinates).

Situation I (type 4, $J_0(\bar{z}) = \varnothing$):

$$v = -\sum_{i_1=1}^{k} y_{i_1}^2 + \sum_{i_2=k+1}^{n-m+1} y_{i_2}^2,$$

$$y_l = 0, l = n - m + 2, \ldots, n.$$

(5.2.26)

Situation II (type 4, $J_0(z) \neq \varnothing$):

$$\delta \cdot v \geqslant -\sum_{i_1=1}^{k} y_{i_1}^2 + \sum_{i_2=k+1}^{c} y_{i_2}^2 - \sum_{i_3=c+1}^{c+d} y_{i_3} + \sum_{i_4=c+d+1}^{c+d+e} y_{i_4},$$

$$y_l \geqslant 0, l = c+1, \ldots, c+d+e,$$

$$y_1 = 0, l = c+d+e+1, \ldots, n,$$

$$c = n - m - |J_0(\bar{z})| + 1, \delta \in \{-1, 1\},$$

$$d + e = |J_0(\bar{z})| - 1, c \neq 0.$$

(5.2.27)

Situation III (type 5): As in situation II, but now $c = 0$; so, the number of active constraints at $\bar{z}$, which is $m + |J_0(\bar{z})|$, equals $n + 1$.

After the description of the feasible set we now have to include the objective function and the fact that we are interested in following a branch of local minimizers for increasing values of the parameter t. Hence, for $t < \bar{t}$ and $t \approx \bar{t}$ we assume that $x(t)$ is a local minimizer for $P(t)$; in particular, we have $(x(t), t) \in \Sigma_{gc}$. Let us denote the objective function in the (y, v) coordinates again by f. We will discuss the situations I, II, III systematically.

Situation I (type 4, $J_0(\bar{z}) = \varnothing$): For simplicity we put $f_i = \partial f(0)/\partial y_i$. From the characterization of points of type 4 (cf. Section 2.5) we have, according to (5.2.26),

$$\Delta := -\sum_{i_1=1}^{k} f_{i_1}^2 + \sum_{i_2=k+1}^{n-m+1} f_{i_2}^2 \neq 0.$$

Since we focus on a branch of local minimizers for $v < 0, v \approx 0$, it is not difficult to see that exactly one of the following possibilities may occur:

$$
\begin{array}{lll}
\text{Ia} & k = 1, \Delta > 0, & \\
\text{Ib} & k = n - m + 1. &
\end{array}
$$

(5.2.28)

We learn from Section 4.4 that the set Σ_{gc} of g.c. points exhibits a quadratic turning point (with respect to t). If we pass the points $\bar{z}$ along Σ_{gc}, then, in cases Ia, Ib, the local minimizer turns into a local maximizer. Moreover, the value

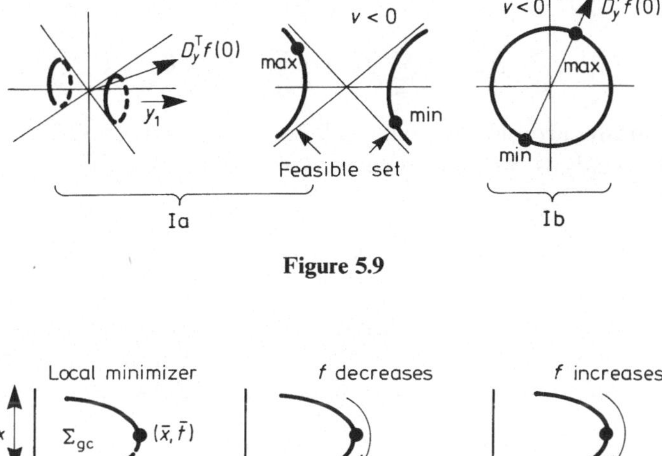

Figure 5.9

Figure 5.10

of f decreases (increases) in cases Ia (Ib). This provides (also in the original coordinates) an easy way for checking whether we have the situation Ia or Ib; see Figure 5.9 and 5.10.

In situation Ia it is possible to jump to another branch of local minimizers. In fact, since the feasible set $M(t)$ is compact, we compute a point on Σ_{gc} beyond the turning point, say $(x_{\max}(t), t)$ with $t < \bar{t}, t$ close to $\bar{t}$; the point $x_{\max}(t)$ is a local maximizer for $P(t)$ and we can start at $x_{\max}(t)$ with a descent method in order to find a local minimizer for $P(t)$, say $\tilde{x}_{\min}(t)$. Since f decreases along Σ_{gc}, the point $\tilde{x}_{\min}(t)$ differs from $x_{\min}(t)$, the local minimizer for $P(t)$ corresponding to $x_{\max}(t)$ on Σ_{gc} (in the neighbourhood of $\bar{z}$); see Figure 5.11.

However, in situation Ib the corresponding component of the feasible set becomes empty, and starting at $x_{\max}(t)$ with a descent method as in Ib would drive us back to the same local minimizer $x_{\min}(t)$!

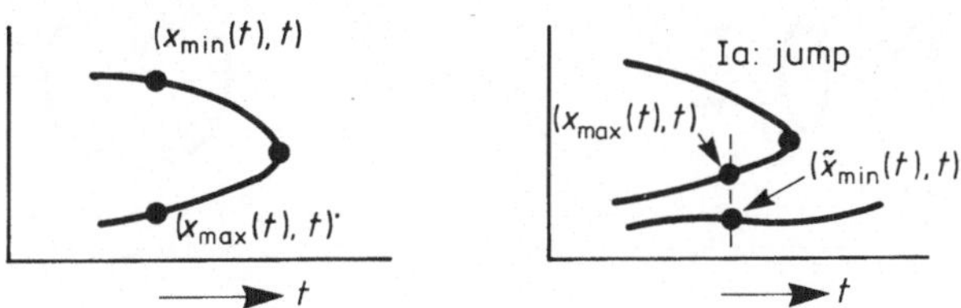

Figure 5.11

Situation II (type 4, $J_0(\bar{z}) \neq \varnothing$): Similarly as in situation I we have

$$\Delta := - \sum_{i_1=1}^{k} f_{i_1}^2 + \sum_{i_2=k+1}^{c} f_{i_2}^2 \neq 0. \qquad (5.2.29)$$

Moreover, locally around the point $\bar{z}$ all linear equalities in (5.2.27) are active. So, for our analysis we can proceed by considering only one inequality constraint:

$$\delta \cdot v \geqslant - \sum_{i_1=1}^{k} y_{i_1}^2 + \sum_{i_2=k+1}^{c} y_{i_2}^2, \qquad \delta \in \{-1, 1\}. \qquad (5.2.30)$$

Since, in addition, we walk on a branch of local minimizers for $v < 0, v \approx 0$, it is again not difficult to see that exactly one of the following possibilities may occur (it is not difficult to see that the number d in (5.2.27) vanishes necessarily):

$$\text{IIa} \qquad \delta = 1, k = 1, \Delta > 0, \qquad (5.2.31)$$

$$\text{IIb} \qquad \delta = -1, k = 0. \qquad (5.2.32)$$

See Figure 5.12 for a picture analogus to Figure 5.9, where $|J_0(\bar{z})| = 1$.

The question whether a jump to another branch of local minimizers is possible or not (by means of a descent method) can be answered completely analogously to situation I. In fact, a jump is possible in situation IIa, whereas it is not so in situation IIb. Moreover, in situation IIb the current component of the feasible set becomes empty.

Situation III (type 5): If $\bar{z} \in \Sigma_{gc}$ is a point of type 5, then exactly $n+1$ constraints are active at $\bar{z}$, and at all points $z \in \Sigma_{gc}, z \neq \bar{z}$, in some neighbourhood O of $\bar{z}$ the number of active constraints equals n. So, in particular, we can conclude from Theorems 2.5.4 and 2.5.5: a branch of local minimizers stops at a point $\bar{z}$ of type 5 if and only if the MFCQ is violated at $\bar{z}$. But then, it is obvious by (5.2.27) that $\delta = -1$ and $d = 0$. In addition, in the latter case the feasible set becomes empty as the parameter t passes the value $\bar{t}$, and a simple way for a jump cannot be proposed.

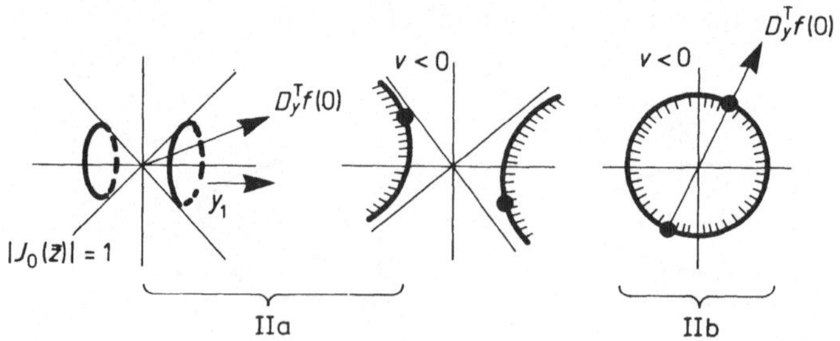

Figure 5.12

We note that the MFCQ is also violated at a g.c. point of type 4. In this case we are in the position to detect the approach of such a point along Σ_{gc} since the norm of the vector of Lagrange multipliers tends to infinity. In the case of a point of type 5 this is different. Although, at a point of type 5, the MFCQ might not be satisfied, the norm of the vector of Lagrange multipliers does *not* tend to infinity as we approach the point. Moreover, if we approach the point $\bar{z}$ along a branch of local minimizers, then $\bar{z}$ itself is also a stationary point, and *at the point $\bar{z}$* the set of Lagrange multipliers is either bounded (MFCQ is satisfied) or not (MFCQ is not satisfied). In order to continue our walk along a branch of local minimizers, we take the Lagrange multipliers (λ, μ) into account and follow a corresponding piecewise smooth curve in the (x, λ, μ, t) space, rather than in the (x, t) space (cf. Section 4.4). The projection of this path into the (λ, μ, t) space and the (x, t) space, respectively takes one of the forms depicted in Figure 5.13.

Summarizing, we can start at $(x^0, 0) \in \Sigma_{gc}^1 \cap \Sigma_{loc}$ (cf. (A2)) using the algorithm PATH I and the algorithm PATH II, respectively. It turned out that, at certain values of the parameter t, the path of local minimizers stops at a point $\bar{z}$, and, in order to follow local minimizers, we have to jump to another branch in Σ_{loc}. In some cases—at a point $\bar{z}$ of type 2, type 3 and type 4 (f decreases locally around $\bar{z}$)—a jump is possible. If the path in $\bar{\Sigma}_{loc}$ stops at a point $\bar{z}$ of type 4 (f increases locally around $\bar{z}$) and of type 5, a simple way for a jump to another connected component in $\bar{\Sigma}_{loc}$ cannot be proposed. Of course, if the path stops at a point of type 5, the MFCQ is violated (cf. Figure 5.13(b)). If the MFCQ is satisfied (cf. Figure 5.13(a)), we can use the proposed algorithms for increasing t in order to follow the path of local minimizers.

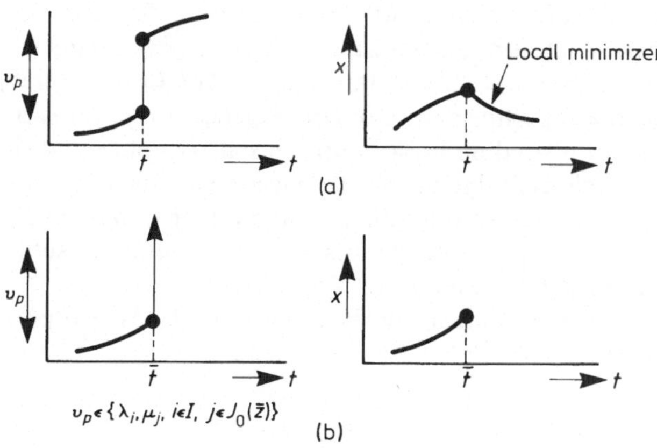

Figure 5.13

Algorithm JUMP I

We start the algorithm with the point $(x^0, 0) \in \Sigma^1_{loc}$, and want to find a discretization (5.1.4) and corresponding local minimizers $x(t_i), i = 1, \ldots, N$.

1. $(x^0(\bar{t}), \lambda^{J^0}(\bar{t}), \mu^{J^0}(\bar{t}))$ is computed by applying PATH II to system (4.4.3). Denote $\bar{z} = (x^{J^0}(\bar{t}), \bar{t})$.

2. If $\bar{z}$ is a point of type 2 with (5.2.20) III, then goto 3; else goto 4.

3. Solve problem (5.2.21) in order to get a quadratic descent direction ξ. Let $\tilde{x}$ be the obtained local minimizer and $J_0 := J_0(\tilde{x}, \bar{t})$. Solve (4.4.3) by PATH II, starting with $\tilde{x}$ at $\bar{t}$. Let $\hat{t}$ be the new point of the discretization. Do $\bar{t} := \hat{t}$ and goto 1.

4. If $\bar{z}$ is a point of type 3 then goto 5; else goto 6.

5. Calculate the tangential direction of descent by (5.2.22). Let $\tilde{x}$ be the obtained local minimizer and $J_0 := J_0(\tilde{x}, \bar{t})$. Solve (4.4.3) by PATH II, starting with $\tilde{x}$ at $\bar{t}$. Let $\hat{t}$ be the new point of the discretization. Do $\bar{t} := \hat{t}$ and goto 1.

6. If $\bar{z}$ is a point of type 4. Then goto 7; else goto 8.

7. If we have situation Ia (cf. (5.2.28)) or IIa (cf. (5.2.30)), then jump (with a standard descent method) to a local minimizer, say $\tilde{x}$; solve (4.4.3) by PATH II, starting with $\tilde{x}$ at $\bar{t}$. Let $\hat{t}$ be the new point of the discretization; do $\bar{t} := \hat{t}$ and goto 1; else EXIT (because we have situation Ib (cf. (5.2.28)) or situation IIb (cf. (5.2.3))).

8. If $\bar{z}$ is a point of type 5 at which the MFCQ is violated then EXIT; else goto 1.

REMARK 5.2.2   The algorithm JUMP I generates a discretization (5.1.4) and corresponding local minimizers $x(t_i), i = 1, \ldots, N$, if only points of type 2, type 3, type 4 (case (5.2.28) and (5.2.31)) and type 5 (MFCQ is satisfied) appear. Of course, we lose the superlinear rate of convergence if a jump is necessary. We proposed a descent procedure for a jump. This also implies that this version of JUMP I is not implementable in the strongest sense, since we need an infinite number of steps to find a local minimizer on another branch in Σ_{loc}. However, a stronger version is possible, because we need only a sufficiently good approximation of the local minimizer $\tilde{x}$.

The following examples illustrate the algorithm JUMP I if points of type 2 and type 3 appear.

Example 5.2.3

(See Example 4.5.2 and Figure 4.10.) The starting point is the local minimizer

Table 5.4

SINGULARITIES	t	x
	0	$-3.714\,192\,64$
(x^1, t_1) type 2	$0.066\,689\,033\,7$	$-3.679\,191\,54$
	0.1	$-3.661\,098\,43$
	0.2	$-3.602\,935\,13$
	0.3	$-3.538\,532\,37$
	0.4	$-3.466\,242\,77$
	0.5	$-3.383\,633\,49$
	0.6	$-3.286\,873\,06$
	0.7	$-3.169\,351\,92$
	0.8	$-3.017\,948\,65$
	0.9	$-2.799\,129\,06$
(x^2, t_2) type 2	$0.967\,155\,465$	$-2.554\,547\,79$
	$\vdots$	$\vdots$
	$0.670\,588$	$-2.554\,547\,79$

Table 5.5

t	x
0.95	$-1.338\,467\,92$
$\vdots$	$\vdots$
1.0	$-1.338\,467\,92$

Table 5.6

SINGULARITIES	t	x
	0	0
	0.1	0.0834
	0.2	0.1595
	0.3	0.2499
	0.4	0.3278
	0.5	0.0266
	0.6	-0.8895
(x^1, t_1) type 3	0.633	-1.212
	0.6	-1.4189
	0.5	-2.1520
	0.4	-2.3815

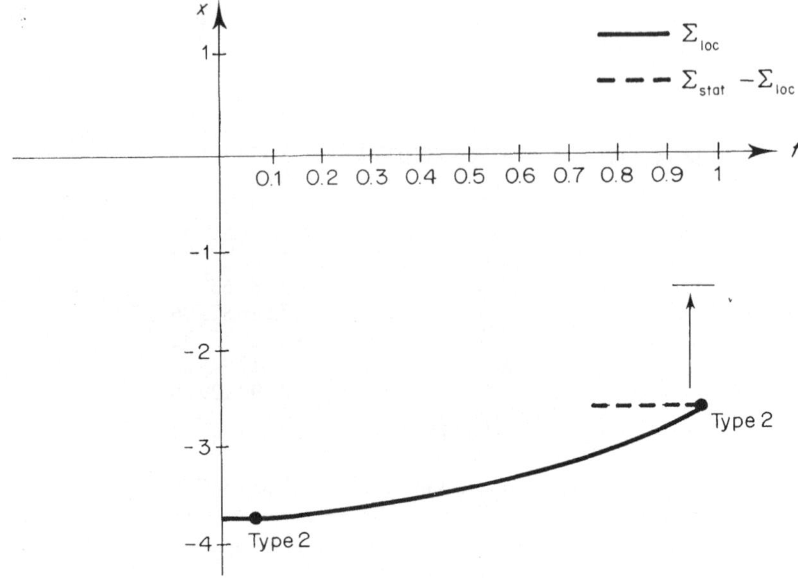

Figure 5.14

Table 5.7

t	x
0.5	-3.1922
0.6	-3.1829
0.7	-3.1402
0.8	-3.1134
0.9	-3.0960
1	-3.084

$x^0 = -3.714\,192\,64$ at $t_0 = 0$. By applying PATH II we obtain Table 5.4 of stationary points (local minimizers for increasing $t \in [0, 10.967\,155\,465))$.

Using a jump close to the turning point (x^2, t_2) of type 2 and applying PATH II once more, we obtain Table 5.5 of local minimizers.

The results are illustrated in Figure 5.14.

As mentioned in Section 4.5.2 with respect to Example 4.5.2 we can find a discretization of $[0, 1]$ and corresponding local minimizers by using PATH I and PATH II, respectively, only if we choose other starting points.

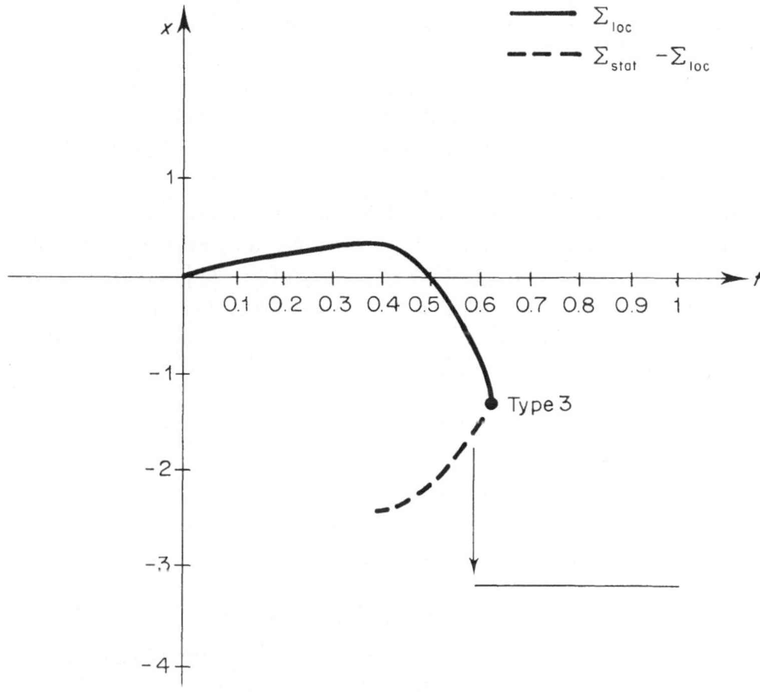

Figure 5.15

Example 5.2.4

(See Example 5.1.1 and Figure 5.1.) The starting point is the local minimizer $x^0 = 0$ at $t_0 = 0$. By applying PATH II we obtain Table 5.6 of stationary points (local minimizers for increasing $t \in [0, 0.633)$).

Using a jump close to the point (x^1, t_1) of type 3 and applying PATH II once more, we obtain Table 5.7 of local minimizers.

The results are illustrated in Figure 5.15.

5.3 JUMPS IN THE CRITICAL SET AND THE ALGORITHM JUMP II

In this section we follow the article by Guddat *et al.* [91]. We assume that (A1), (A2) and (A3′) are satisfied; (A3′) is an extension of (A3) and is introduced below. The local structure of Σ_{gc} is depicted in Figure 2.16. We modify the aim of our investigation in the following way.

Find a numerical description of as many connected components of Σ_{gc} as possible by using the algorithm PATH III and jumps in Σ_{gc}. Of course, if we can follow numerically all the connected components of Σ_{gc} (we will see that we have a finite numer of connected components, cf. Theorem 5.3.1), we have surely found a discretization (5.1.4) of the interval $[0, 1]$ and corresponding g.c. points $x(t_i), i = 1, \ldots, N$.

Since we cannot show that we find all connected components surely, the aim was formulated in the above sense. We consider the parametric optimization problem $P(t), t \in [t_A, t_B]$, where $[0, 1] \subsetneqq [t_A, t_B]$ and $[t_A, t_B]$ is a sufficiently large interval. 'Sufficiently large' means that we assume that there is no turning point $\bar{z} = (\bar{x}, \bar{t}) \in \Sigma_{gc}$ with $\bar{t} \in (-\infty, t_A)$ and $\bar{t} \in (t_B, +\infty)$, respectively. We modify (A3) for our investigation by:

(A3′) For all $t \in [t_A, t_B], M(t)$ is non-empty and there exists a compact set C containing $M(t)$.

We note that the extension (A3′) of (A3) is only of technical character.

By $\Sigma_{gc} | [t_A, t_B]$ we denote the set of g.c. points restricted to the interval $[t_A, t_B]$, i.e.

$$\Sigma_{gc} | [t_A, t_B] = \{(x, t) \in \mathbb{R}^n \times [t_A, t_B] \mid x \text{ is a g.c. point for } P(t)\}.$$

Then, by using the properties of the class $\mathscr{F}^{**}$, (A1) and the compactness of $\Sigma_{gc} | [t_A, t_B]$ (cf. (A2) and (A3′)), we have

THEOREM 5.3.1   Assume that (A1), (A2) and (A3′) are satisfied. Then, the number of connected components of $\Sigma_{gc} | [t_A, t_B]$ is finite.

As noted in Section 5.1, an estimation of this number is still not known.

We consider the connected component $C = C(x^0, t_0)$ in Σ_{gc} introduced in Section 4.4 and we restrict this connected component also to the interval $[t_A, t_B]$ denoted by $C | [t_A, t_B]$. Using the algorithm PATH III with the starting point $(x^0, t_0) \in \Sigma_{gc}^1$ we assume that $C | [t_A, t_B]$ is completely described numerically and that we have found a discretization (sufficiently fine) of the interval $[t_A, t_B]$:

$$t_{-q_1} < \cdots < t_0 < \cdots < t_i < t_{i+1} < \cdots < t_{r_1}$$

and the corresponding set $C | [t = t_i]$ of g.c. points $i \in \{-q_1, \ldots, 0, \ldots, r_1\}$, where

$$t_{-q_1} \in [t_A, t_0), \qquad t_{r_1} \in (t_0, t_B]$$

$$C | [t = t_i] := \{(x, t_i) \in \mathbb{R}^n \times \mathbb{R} \mid (x, t_i) \in C\}. \tag{5.3.1}$$

Furthermore, we see that the set $C | [t = t_i], i \in \{-q_1, \ldots, 0, \ldots, r_1\}$, contains a finite number of g.c. points. The algorithm PATH III could easily be organized

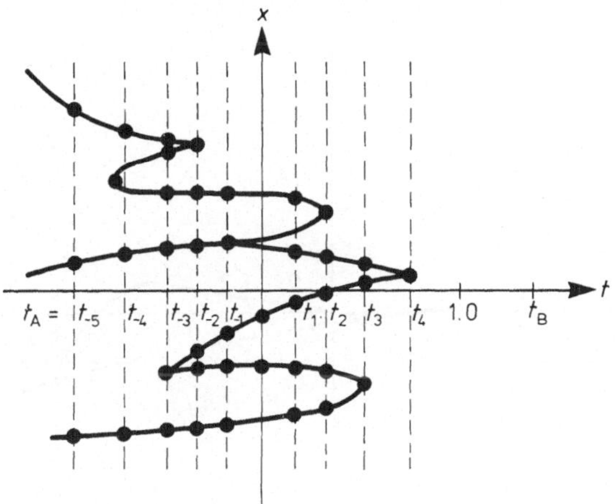

Figure 5.16

in such a way that all singularities of $C|[t_A, t_B]$ are included (the step length is suitably chosen). This situation is depicted in Figure 5.16. In this figure $t_A = t_{-q_1}$ and $t_{r_1} < t_B$ and even $t_{r_1} < 1$.

For each $t_i, i \in \{-q_1, \ldots, 0, \ldots, r_1\}$, we can now easily compute the points $\hat{x}(t_i)$ and $\bar{x}(t_i)$ with the following properties:

$$f(\hat{x}(t_i), t_i) \leqslant f(x, t_i) \qquad \text{for all } (x, t_i) \in C/[t = t_i], \qquad (5.3.2)$$

and

$$f(\bar{x}(t_i), t_i) \geqslant f(x, t_i) \qquad \text{for all } (x, t_i) \in C/[t = t_i]. \qquad (5.3.3)$$

Of course, $\hat{x}(t_i)$ and $\bar{x}(t_i)$, respectively, need not be unique. Now we have the following possibilities to jump to another connected component in Σ_{gc}.

Situation 1: There exists an

$$i_0 \in \{-q_1, \ldots, 0, \ldots, r_1\} \quad \text{with} \quad (\hat{x}(t_{i_0}), t_{i_0}) \in \Sigma_{gc}^1 \backslash \Sigma_{stat}. \qquad (5.3.4)$$

Then we can compute a direction $\xi \in \mathbb{R}^n, \xi \neq 0$, of descent in the normal way using the condition:

$$\begin{aligned} D_x f(\hat{x}(t_{i_0}), t_{i_0})\xi < 0, \qquad & D_x h_i(\hat{x}(t_{i_0}), t_{i_0})\xi = 0, \qquad i \in I, \\ D_x g_j(\hat{x}(t_{i_0}), t_{i_0})\xi \leqslant 0, \qquad & j \in J_0(\hat{x}(t_{i_0}), t_{i_0}). \end{aligned} \qquad (5.3.5)$$

(For $(\bar{x}(t_{i_0}), t_{i_0}) \in \Sigma_{gc}^1 \backslash \Sigma_{stat}$, we compute a feasible direction of descent for $-f(x, t_{i_0})$.)

The following example shows that situation 1 could appear.

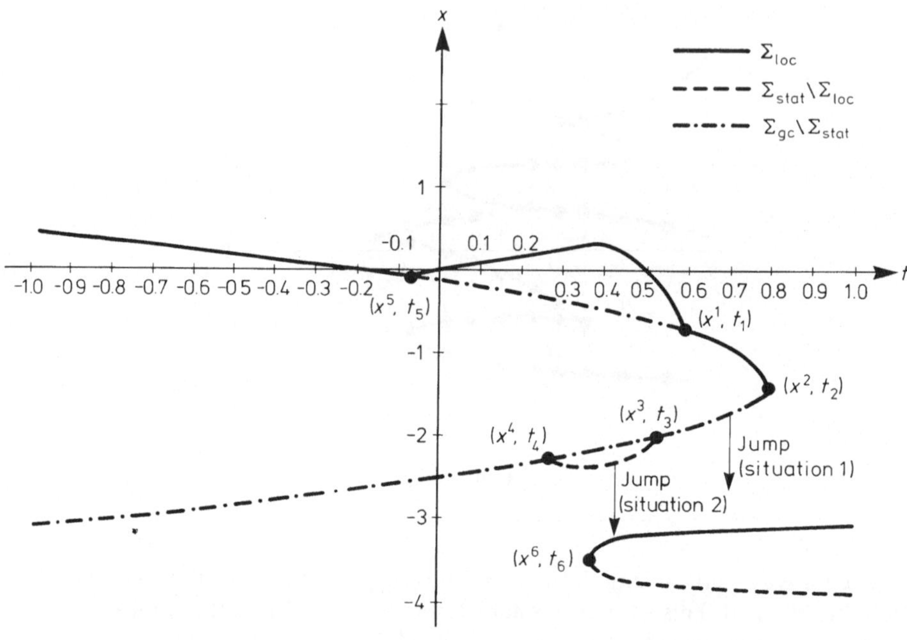

Figure 5.17

Example 5.3.1

(See Figure 5.17.)

$$P(t): \qquad \min\{f(x,t)\,|\,x\in M(t)\}, \qquad t\in[-1,1]$$

where $f(x,t)$ is defined by (5.1.4) and

$$M(t):= \{x\in\mathbb{R}\,|\,g(x,t)\leqslant 0\},$$
$$g(x,t):= -(x+1)^2 - t + 0.8.$$

Σ_{gc} is depicted in Figure 5.17.

We note that the check (5.3.4) is easy to realize: we have to consider the corresponding Lagrange multipliers $\mu_j(t_{i_0})$, $j\in J_0(\hat{x}(t_{i_0}), t_{i_0})$. They have to be different from zero and there exists a $j^*\in J_0(\hat{x}(t_{i_0}), t_{i_0})$ with $\hat{\mu}_{j^*}(t_{i_0}) < 0$.

Situation 2: There exists an

$$i_0\in\{-q_1,\ldots,0,\ldots,r_1\} \quad \text{with} \quad (\hat{x}(t_{i_0}), t_{i_0})\in\Sigma^1_{\text{stat}}\setminus\Sigma_{\text{loc}}. \tag{5.3.6}$$

Then, we cannot compute a direction of descent by using (5.3.5) but there exists a direction $\xi \in \mathbb{R}^n, \xi \neq 0$, of descent with the property

$$\min\{\xi^T D_x^2 L(\hat{x}(t_{i_0}), t_{i_0})\xi \mid \|\xi\| = 1, \qquad D_x h_i(\hat{x}(t_{i_0}), t_{i_0})\xi = 0, \qquad i \in I,$$
$$D_x g_j(\hat{x}(t_{i_0}), t_{i_0})\xi \leqslant 0, \qquad j \in J_0(\hat{x}(t_{i_0}), t_{i_0})\} < 0\}. \tag{5.3.7}$$

This follows immediately from the property (5.3.6), since $D_x^2 L(\hat{x}(t_{i_0}), t_{i_0})/T(\hat{x}(t_{i_0}), t_{i_0})$ is non-singular and not positive definite. (For $(\bar{x}(t_{i_0}), t_{i_0}) \in \Sigma_{\text{stat}}^1 \setminus \Sigma_{\text{loc}}$, we compute a feasible direction of descent for $-f(x, t_{i_0})$.)

Example 5.1.1 shows that this situation could appear, too (cf. Figure 5.17). We note that the check of (5.3.6) is not so easy. We refer to Richter [178] to observe the smallest eigenvalue of the matrix $D_x^2 L(x(t), t)/T(x(t), t)$ by numerical tracing of a curve of critical points.

Furthermore, the problem in (5.3.7) is a non-convex quadratic optimization problem if we choose $\|\cdot\|$ in a suitable way. To find a direction of descent we have to compute a global minimizer in the extreme case (cf. Remark 5.2.1, too).

For the next situations we assume for simplicity of explanation that the turning points on the connected component $C = C(x^0, 0)$ are computed exactly (for the algorithm, the exact knowledge is not necessary, cf. Section 5.2 and the algorithm JUMP 1 in this section).

Situation 3: There exists an

$$i_0 \in \{-q_1, \ldots, 0, \ldots, r_1\} \text{ with } (\mathring{x}(t_{i_0}), t_{i_0}) \in \bar{\Sigma}_{\text{loc}}^2, \tag{5.3.8}$$

and $(\hat{x}(t_{i_0}), t_{i_0})$ is a turning point in Σ_{stat}. Then ξ with the property (5.2.21) is a feasible direction of descent for $f(x, t_{i_0})$ (cf. Remark 5.2.1). For details we refer to Section 5.2. (For $(\bar{x}(t_{i_0}), t_{i_0}) \in \bar{\Sigma}_{\text{loc}}^2$, we have to replace $f(x, t_{i_0})$ by $-f(x, t_{i_0})$ in (5.2.21).)

Situation 4: There exists an

$$i_0 \in \{-q_1, \ldots, 0, \ldots, r_1\} \text{ with } (\hat{x}(t_{i_0}), t_{i_0}) \in \bar{\Sigma}_{\text{loc}}^3. \tag{5.3.9}$$

Then, $u(t)$ defined by (5.2.22) tends to a direction of descent if t tends to t_{i_0} (for details, cf. Section 5.2.). (For $(\bar{x}(t_{i_0}), t_{i_0}) \in \Sigma_{\text{loc}}^3$, we have to choose $-u(t)$ in (5.2.22).)

Situation 5: There exists an

$$i_0 \in \{-q_1, \ldots, 0, \ldots, r_1\} \quad \text{with} \quad (\hat{x}(t_{i_0}), t_{i_0}) \in \bar{\Sigma}_{\text{loc}}^4, \tag{5.3.10}$$

and $(\hat{x}(t_{i_0}), t_{i_0})$ satisfies either (5.2.28) Ia or (5.2.31) IIa. Then, we compute a point on Σ_{gc} beyond the turning point $(\hat{x}(t_{i_0}), t_{i_0})$, say $(x_{\max}(t), t)$; the point $x_{\max}(t)$ is a local maximizer for $P(t)$ and we can start with the descent method at $x_{\max}(t)$ in order to jump to another connected component in Σ_{gc}. (For $(\bar{x}(t_{i_0}), t_{i_0}) \in \bar{\Sigma}_{\text{loc}}^4$ and $(\bar{x}(t_{i_0}), t_{i_0})$ satisfying (5.2.29) Ib or (5.2.3) IIb, we can jump analogously with a descent method for $-f(x, t_{i_0})$.)

From the analysis in Section 5.2 it follows that the situations $i, i \in \{3, 4, 5\}$, could appear. For all further possible situations, the reader can easily find that we still do not have any proposals for jumps.

We also get jumps to another connected component in Σ_{gc} at the points $(\bar{x}(t_{i_0}), t_{i_0})$ with (5.3.3) in exactly the same situations, but we have to construct a direction of descent for the function $-f(x, t_{i_0})$ with respect to $M(t_{i_0})$.

From this consideration we derive the algorithm.

Algorithm JUMP II

For simplicity we restrict the algorithm to the computation of at most seven connected components. The extension to more than seven connected components can be done by using the same approach described in the steps 10–15.

We start the algorithm with the starting point $(x^0, t_0) \in \Sigma^1_{gc}$.

1. $k := 1, t^1_- := t_A, t^1_+ := t_B, x^1_0 := x^0, t^1_0 := t_0$.

2. Starting point (x^k_0, t^k_0). Apply algorithm PATH III in order to describe a connected component $C_k | [t^k_-, t^k_+]$ numerically.

3. If $k = 7$ then STOP.

4. Let $t^k_{-q_k} < \cdots < t^k_0 < \cdots < t^k_1 < t^k_{i+1} < \cdots < t^k_{r_k}$ be the found discretization of the interval $[t^k_-, t^k_+]$, organized in such a way that all singularities in $C_k | [t^k_-, t^k_+]$ are included.

5. If $k \leqslant 3$, then put in $\hat{x}^k(t^k_i), \bar{x}^k(t^k_i)$, the corresponding points with the properties (5.3.2), (5.3.3), respectively, $i = -q_k, \ldots, 0, \ldots, r_k$.

6. $k := k + 1$.

7. If $k = 2$ then $t^2_- := t_A, t^2_+ := t_B, \hat{x}(t_i) := \hat{x}^1(t^1_i)$, for $i = -q_1, \ldots, 0, \ldots, r_1$, goto 16.

8. If $k = 3$ then $t^3_- := t_A, t^3_+ := t_B, \hat{x}(t_i) := \bar{x}^1(t^1_i)$, for $i = -q_1, \ldots, 0, \ldots, r_1$, $f := -f$, goto 16.

9. If $k > 4$ then goto 13.

10. If $t^2_{r_2} \leqslant \max(t^1_{r_1}, t^3_{r_3})$ or $t^3_{-q_3} \geqslant \min(t^1_{-q_1}, t^2_{-q_2})$ then EXIT.

11. $t^*_+ := \max(t^1_{r_1}, t^3_{r_3}), t^*_- := \min(t^1_{-q_1}, t^2_{-q_2})$.

12. $t^4_- := t_A, t^4_+ := t_B, \hat{x}(t_i) := \hat{x}^2(t^2_i)$ for $i = -q_2, \ldots, 0, \ldots, r_2$, goto 16.

13. If $k = 5$ then $t^5_- := t^*_+, t^5_+ := t_B, \hat{x}(t_i) := \bar{x}^2(t^2_i)$ for all $i \in \{\{-q_2, \ldots, 0, \ldots, r_2\} | t^2_i \geqslant t^*_+\}, f := -f$, goto 16.

14. If $k = 6$ then $t_-^6 := t_{-q_3}^3, t_+^6 := t_-^*, \hat{x}(t_i) := \hat{x}^3(t_i^3)$ for all $i \in \{\{-q_3, \ldots, 0, \ldots, r_3\} \mid t_i^3 \leqslant t_-^*\}$, goto 16.

15. If $k = 7$ then $t_-^7 := t_A, t_+^7 := t_B, \hat{x}(t_i) := \bar{x}^3(t_i^3)$ for $i = -q_3, \ldots, 0, \ldots, r_3, f := -f$.

16. Look for the first i such that one of the following conditions holds (choose the first condition that is satisfied):

   (C1) $(\hat{x}(t_i), t_i) \in \Sigma_{gc}^1 \setminus \Sigma_{stat}$,

   (C2) $(\hat{x}(t_i), t_i) \in \Sigma_{stat}^1 \setminus \Sigma_{loc}$,

   (C3) $(\hat{x}(t_i), t_i) \in \bar{\Sigma}_{loc}^2$ and $(\hat{x}(t_i), t_i)$ is a turning point,

   (C4) $(\hat{x}(t_i), t_i) \in \bar{\Sigma}_{loc}^3$,

   (C5) $(\hat{x}(t_i), t_i) \in \bar{\Sigma}_{loc}^4$ and $(\hat{x}(t_i), t_i)$ satisfies either (5.2.28) Ia or (5.2.3) IIa (resp. (5.2.28) Ib or (5.2.3) IIb in the case when f was replaced by $-f$).

17. If (C1) holds, then calculate a direction $\xi \in \mathbb{R}^n, \xi \neq 0$, of descent by using the condition (5.3.5), jump to a new point, say $\tilde{x}$, on another connected component in $\Sigma_{gc}, x_0^k := \tilde{x}, t_0^k := t_i$, goto 2.

18. If (C2) holds, then calculate a direction $\xi \in \mathbb{R}^n, \xi \neq 0$, of descent by solving the problem (5.3.7), jump to a new point, say $\tilde{x}$, on another connected component in $\Sigma_{gc}, x_0^k := \tilde{x}, t_0^k := t_i$, goto 2.

19. If (C3) holds, then calculate a direction $\xi \in \mathbb{R}^n, \xi \neq 0$, of descent by solving the problem (5.2.21), jump to a new point, say $\tilde{x}$, on another connected component in $\Sigma_{gc}, x_0^k := \tilde{x}, t_0^k := t_i$, goto 2.

20. If (C4) holds, then calculate a direction $\xi \in \mathbb{R}^n, \xi \neq 0$, of descent by using condition (5.2.22) ($-\xi$ in the case when f was replaced by $-f$), jump to a new point, say $\tilde{x}$, on another connected component in $\Sigma_{gc}, x_0^k := \tilde{x}, t_0^k := t_i$, goto 2.

21. If (C5) holds, then calculate a point on $\Sigma_{gc} \mid [t_A, t_B]$ beyond the turning point $(\hat{x}(t_i), t_i)$, say $(x_{max}(t), t)$, jump from $x_{max}(t)$, with a descent method, to a new point, say $\tilde{x}$, on another connected component in $\Sigma_{gc}, x_0^k := \tilde{x}, t_0^k := t_i$, goto 2.

Figure 5.18 shows the main ideas on which algorithm JUMP II is based. The following examples illustrate algorithm JUMP II.

Example 5.3.2

(See Figure 5.17.) The starting point for the algorithm JUMP II is the local minimizer $x^0 = 0$ at $t_0 = 0$. We get the numerical description shown in Table 5.8 of the connected component $C(x^0, 0) \mid [-1, 1]$ by using the algorithm PATH III.

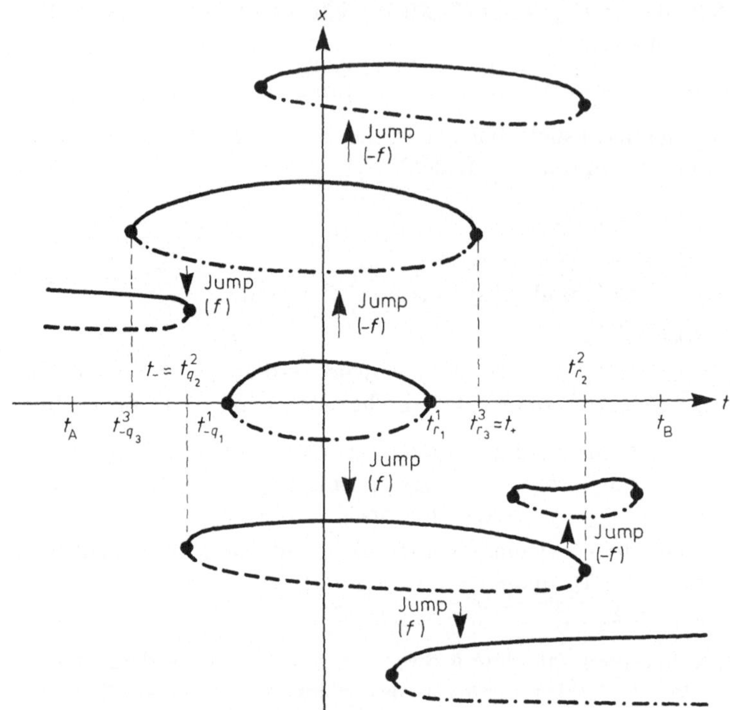

Figure 5.18

Table 5.8

SINGULARITIES	t	x
	0	0
	0.1	0.0834
	0.2	0.1595
	0.3	0.2499
	0.4	0.3278
	0.5	0.0266
(x^1, t_1) type 2	0.596	−0.7315
	0.6	−0.7386
	0.7	−0.9122
(x^2, t_2) type 3	0.8	−1.4078
	0.7	−1.7534
	0.6	−1.9196

Table 5.8 *Contd.*

SINGULARITIES	t	x
(x^3, t_3) type 2	0.525	-2.0318
	0.5	-2.1520
	0.4	-2.3815
	0.3	-2.3047
(x^4, t_4) type 2	0.262	-2.3111
	0.2	-2.3708
	0.1	-2.4480
	0	-2.5260
	-0.1	-2.5982
	-0.2	-2.6666
	-0.3	-2.7317
	-0.4	-2.7939
	-0.5	-2.8537
	-0.6	-2.9109
	-0.7	-2.9663
	-0.8	-3.0199
	-0.9	-3.0718
	-1.0	-3.1222
(x^1, t_1) type 2	0.596	-0.7315
	0.5	-0.6109
	0.4	-0.4778
	0.3	-0.3965
	0.2	-0.2846
	0.1	-0.2155
	0	-0.1406
(x^5, t_5) type 2	-0.076	-0.0896
	0.1	-0.0684
	-0.2	0
	-0.3	0.0651
	-0.4	0.1273
	-0.5	0.1869
	-0.6	0.2443
	-0.7	0.2997
	-0.8	0.3532
	-0.9	0.4051
	-1.0	0.4552
(x^3, t_3) type 2	0.525	-2.0318
	0.5	-2.0531
	0.4	-2.1759
	0.3	-2.2745
(x^4, t_4) type 2	0.262	-2.3708

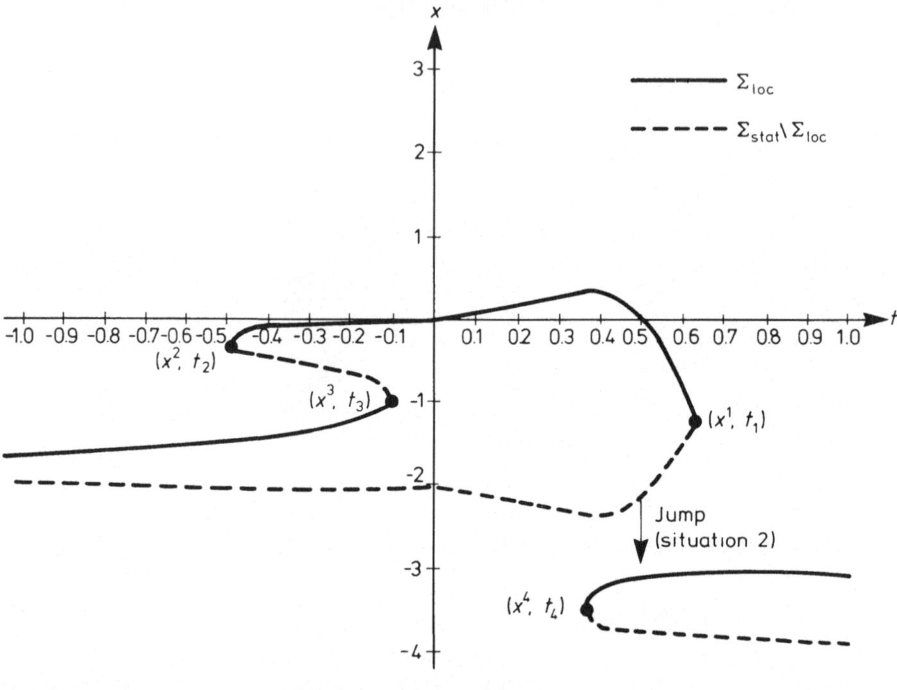

Figure 5.19

Example 5.3.3

(See Figure 5.19.)

$$P(t): \qquad\qquad \min\{f(x,t)\,|\,x\in\mathbb{R}^n\}, \qquad t\in[-1,1]$$

where $f(x,t)$ is defined by (5.1.5). The starting point for the algorithm **JUMP II** is the local minimizer $x^0 = 0$ at t_0. We get the numerical description of $C(x^0, t_0)|[-1,1]$ shown in Table 5.9.

Table 5.9

SINGULARITIES	t	x
	0	0
	0.1	0.0834
	0.2	0.1595
	0.3	0.2499
	0.4	0.3279
	0.5	0.0266
	0.6	-0.8895
(x^1, t_1) type 3	0.628	-1.2346

Table 5.9 *Contd.*

SINGULARITIES	t	x
	0.6	-1.4189
	0.5	-2.1520
	0.4	-2.3814
	0.3	-2.3047
	0.2	-2.2118
	0.1	-2.1218
	0	-2.0288
	-0.15	-2.0542
	-0.25	-2.0555
	-0.35	-2.0449
	-0.45	-2.0300
	-0.53	-2.0148
	-0.63	-1.9998
	-0.73	-1.9852
	-0.82	-1.9852
	-0.92	-1.9571
	-1.02	-1.9434
	0	0
	-0.2	-0.0830
	-0.3	-0.1263
	-0.4	-0.1263
	-0.45	-0.2190
(x^2, t_2) type 3	-0.49	-0.2713
	-0.493	-0.3528
	-0.44	-0.4325
	-0.37	-0.5057
	-0.28	-0.5818
	-0.23	-0.6518
	-0.17	-0.7294
	-0.13	-0.8156
	-0.107	-0.9095
(x^3, t_3) type 3	-0.1008	-1.0081
	-0.115	-1.1081
	-0.14	-1.1766
	-0.18	-1.2465
	-0.24	-1.3153
	-0.31	-1.3771
	-0.4	-1.4309
	-0.48	-1.4746
	-0.58	-1.5100
	-0.68	-1.5445
	-0.78	-1.5727
	-0.87	-1.5980
	-0.98	-1.6209
	-1.07	-1.6422

For example, $(-2.152, 0.5) \in \Sigma^1_{\text{stat}} \setminus \Sigma_{\text{loc}}$ (situation 2), and we jump to $(-3.1922, 0.5) \in \Sigma^1_{\text{loc}}$ on the second connected component (Table 5.10).

Table 5.10

SINGULARITIES	t	x
	0.5	-3.1922
	0.6	-3.1829
	0.7	-3.1402
	0.8	-3.1134
	0.9	-3.0960
	1	-3.0841
	0.5	-3.1922
	0.4	-3.2917
(x^3, t_3) type 3	0.371	-3.5096
	0.4	-3.6493
	0.5	-3.7685
	0.6	-3.7995
	0.7	-3.8414
	0.8	-3.8655
	0.9	-3.8808
	1	-3.8912

6

Applications

6.1 PRELIMINARY OUTLINE

In this chapter we consider three applications of our approach: globally convergent algorithms for nonlinear optimization problems (Section 6.2), global optimization (Section 6.3) and multi-objective optimization (Section 6.4). It will be shown that pathfollowing methods with jumps can be useful for the considered applications but not successful in every case. This is no surprise, since we do not have jumps in all situations that can appear in the set Σ_{gc} under the assumption (A1) (cf. Chapter 5). The following problems remain open:

(i) Find an actually globally convergent algorithm for a non-convex optimization problem with an arbitrarily chosen non-feasible starting point without any additional information.

(ii) Find a deterministic approach to solve surely the problem of global optimization.

(iii) Find surely a goal realizer (cf. Definition 6.4.1) for the multi-objective optimization problem if the goal (which expresses the current wishes of the decision-maker in a dialogue procedure) has been a realistic one and give an answer to the question whether the goal has been a realistic one or not.

To solve these problems surely we have to find a numerical description of all connected components in Σ_{gc} in the worst case, but for this we may need jumps in all possible situations. From this point of view the three problems have a common difficulty. This will be shown clearly in this chapter and we will get a deeper insight into these three fundamental problems. With respect to (i) and (ii) our approach could be helpful if other proposals are not successful. In multi-objective optimization it will be shown that the use of pathfollowing methods with jumps has some advantages in comparison with the traditional way, on the one hand. On the other hand, different parametrizations can be compared from the following point of view: what kind of singularities could appear?

6.2 ON GLOBALLY CONVERGENT ALGORITHMS

We consider the following nonlinear optimization problem

(P) $$\min\{f(x)\,|\,x\in M\}\tag{6.2.1}$$

where

$$M:=\{x\in\mathbb{R}^n\,|\,h_i(x)=0,\,i\in I,\,g_j(x)\leqslant 0,\,j\in J\},\tag{6.2.2}$$

with

$$I:=\{1,\ldots,m\},\qquad m<n,\qquad J:=\{1,\ldots,s\}$$

and f, h_i, $g_j\in C^2(\mathbb{R}^n,\mathbb{R})$, $i\in I$, $j\in J$.

From time to time we need a higher degree of differentiability. We follow here Guddat *et al.* [93], Gfrerer *et al.* [70, 71] and Guddat [82].

We remind the reader of the following notions from the theory of point-to-set maps (cf. e.g. Bank *et al.* [12]). A point-to-set map $\Gamma:T\to 2^X$ (T,X are metric spaces) is said to be *upper-semicontinuous* according to Berge (u.s.c.-B) at a point $t^0\in T$ if for each open set Ω with $\Gamma(t^0)\subset\Omega$ there exists a $\delta=\delta(\Omega)>0$ such that $\Gamma(t)\subset\Omega$ for all t in a δ neighbourhood $V_\delta(t^0)$ of t^0.

Γ is said to be *lower-semicontinuous* according to Berge (l.s.c.-B) if for each open set Ω with $\Omega\cap\Gamma(t^0)\neq\varnothing$ there exists a $\delta=\delta(\Omega)$ such that $\Omega\cap\Gamma(t)\neq\varnothing$ for all $t\in V_\delta(t^0)$.

As introduced in Section 1.2 we use the well known approach of embedding. Choose a one-parametric optimization problem

$P(t)$: $$\min\{f(x,t)\,|\,x\in M(t)\},\qquad t\in[0,1],\tag{6.2.3}$$

where

$$M(t):=\{x\in\mathbb{R}^n\,|\,h_i(x,t)=0,\,i\in I,\,g_j(x,t)\leqslant 0,\,j\in J\},$$
$$I:=\{1,\ldots,m\},\qquad m<n,\qquad J:=\{1,\ldots,s\}$$

with at least the following properties:

(B1) A local minimizer for $P(0)$ is known (and the corresponding Lagrange multipliers are easy to compute).

(B2) $P(1)\equiv(\text{P})$.

(B3) $\psi(t)\neq\varnothing$ for all $t\in[0,1]$, where $\psi(t)$ denotes the set of all global minimizers. From time to time we have to substitute (B3) by a stronger condition.

(B3$'$) For all $t\in[0, 1]$, $M(t)$ is non-empty and there exists a compact set C such that $M(t)\subset C$.

This is a minimum of properties for finding a discretization of $[0, 1]$:

$$0 = t_0 < \cdots < t_i < \cdots < t_N = 1 \qquad (6.2.4)$$

and corresponding

(A) local minimizers $x(t_i)$, $i = 1, \ldots, N$,

or

(B) stationary points $x(t_i)$, $i = 1, \ldots, N$,

or

(C) g.c. points $x(t_i)$, $i = 1, \ldots, N$.

For this application it is very important to achieve $t = 1$. In the case when the process is stopped at a point $(\bar{x}, \bar{t})\in\Sigma_{gc}$, $\bar{t} < 1$ and $\bar{x}\notin M$, we have not gained anything.

Now we consider some properties for $P(t)$ that are helpful to find a discretization (6.2.4):

(B4) $(f, H, G)\in\mathscr{F}^{**}$

or

(B4$'$) zero is a regular value of $\mathscr{H}$ defined by (4.1.1).

For a solution algorithm for the problem (P) based on embedding, the simplest but the most restrictive assumptions are (E1), (V1), (V2) and (V3) from Chapter 3 with $t_A = 0$ and $t_B = 1$. Then we have

THEOREM 6.2.1   Assume (E1) and (V1)–(V5) with $t_A = 0$ and $t_B = 1$, (B1) and (B2). Then the algorithm based on PATH I converges at least superlinearly to the local minimizer $x(1)$ for the problem (P).

Proof   Using Theorem 3.4.1 the algorithm PATH I generates a point x^N with $\|x^N - x(1)\| < \varepsilon$ ($\varepsilon > 0$ sufficiently small) in a finite number of steps. Then, x^N lies in the domain of convergence for $x(1)$ with respect to $P^{J_0(x(1),1)}(1)$. Remark 3.2.1 completes the proof. $\square$

Of course, the actual rate of convergence depends on the chosen locally convergent algorithm.

In the case when (B4$'$) is satisfied, we know about the structure of Σ_{stat} (PC^1 paths and PC^1 loops; for details cf. Section 2.6) and we can try to use the algorithm PATH II. The following theorem includes sufficient conditions for this algorithm to be successful.

THEOREM 6.2.2   Assume that the following conditions are satisfied: (B1), (B2), (B3′), (B4′) and also

(B5) For each t in some neighbourhood of zero, there is a unique solution $(x(t), \lambda(t), \mu(t), t)$ of $\mathcal{H}(x, \lambda, \mu, t) = 0$ ($\mathcal{H}$ is defined by (4.1.1)).

(B6) The MFCQ is satisfied for all $x \in M(t)$ for all $t \in [0, 1]$.

Then the algorithm based on PATH II converges at least superlinearly to a stationary point of the problem (P).

Proof   We only have to show (cf. the proof of Theorem 6.2.1) that there exists a PC^1 path S in $\mathcal{H}^{-1}(0)$ that connects $(x^0, \lambda^0, \mu^0, 0)$ with some point $(x^*, \lambda^*, \mu^*, 1)$, where (x^0, λ^0, μ^0) is known by asumption (B1). Of course, by (B2), then x^* is a stationary point for (P). If we have shown the existence of S, we can apply the algorithm PATH II and the Theorem 4.3.1 completes the proof. (B4′) and (B5) imply that there exists a function $\theta: \mathbb{R} \to \mathbb{R}^n \times \mathbb{R}^m \times \mathbb{R}^s \times \mathbb{R}$, $\theta(s) = (x(s), \lambda(s), \mu(s), t(s))$ with $\theta(0) = (x^0, \lambda^0, \mu^0, 0)$ and $t(s) > 0$ for each $s > 0$ (cf. Section 4.2). Suppose that $t(s) < 0$ for each $s > 0$. Since $\|\theta(s)\| \to \infty$ for $s \to \infty$ (cf. Kojima and Hirabayashi [135]) and $t(s) \in (0, 1)$, $x(s) \in s$ for $s > 0$ (cf. (B3′)), there exists a sequence $\{s^i\}$ with $s^i \to \infty$, $t(s^i) \to \tilde{t} \in [0, 1]$, $x(s^i) \to \tilde{x}$ and $\|\lambda(s^i), \mu(s^i)\| \to \infty$. Since $\mu_j(s^i) = g_j(x(s^i), t(s^i))$ (cf. Section 4.2) tends to $g_j(\tilde{x}, \tilde{t})$ for $j \in J$, the sequence $\|\lambda(s^i), \mu(s^i)\|$ tends to infinity, too. (B6) implies that the point-to-set mapping defined by

$$U: \mathbb{R}^n \times \mathbb{R} \to 2^{\mathbb{R}^m \times \mathbb{R}^s},$$

where

$$U(x, t) := \{(\lambda, \mu) \in \mathbb{R}^m \times \mathbb{R}^s \,|\, (\lambda, \mu) \text{ the vector of Lagrange parameters}$$
$$\text{corresponding to } (x, t)\}$$

is u.s.c. at $(\tilde{x}, \tilde{t})$ and $U(x, t)$ is bounded (cf. Theorem 2.4.1), which contradicts $\|\lambda(s^i), \mu(s^i)\| \to \infty$. □

The special embedding $P^4(t)$ provides an example of the realization of Theorem 6.2.2.

REMARK 6.2.1   We note that we are able to find a stationary point by path-following methods only. That is the reason for looking carefully at the assumptions of this theorem. The embedding $P^4(t)$ shows us that only the assumptions (B5) and (B6) could be restrictive.

From Theorem 2.3.2 we learn that the assumptions (B3′) and (B6) imply that $M(t_1)$ is homeomorphic to $M(t_2)$ for all $t_1, t_2 \in [0, 1]$; in particular $M(0)$ is homeomorphic to $M(1)$. (B5) means that only one KKT point exists for $P(0)$.

Then, by using Corollary 4.1 in [87], it follows that $M = M(1)$ is homeomorphic to the unit ball. For the problem (P) defined by (6.2.1) this means that the feasible set M is 'convex-like' in this sense. In particular, a non-connected feasible set M is not possible. From this point of view, we note that the assumptions (B5) in connection with (B6) restrict the class of optimization problems considerably.

REMARK 6.2.2 It is easy to see that we can replace (B6) by the somewhat weaker assumption

(B6′) The MFCQ is satisfied for all x with $(x,t)\in\bar{\Sigma}_{\text{stat}}|[0,1]$.

In the following we drop the assumption (B5) and assume (B4) instead of (B4′). Then, jumps from one connected component in $\bar{\Sigma}_{\text{loc}}$ to another are possible (cf. Section 5.2).

THEOREM 6.2.3 Assume that the following conditions are satisfied: (B1), (B2), (B3′), (B4) and (B6). Then the algorithm based on JUMP I converges to a local minimizer for (P).

Proof Since (B6) is satisfied, we can find a direction of descent if a point of type 3 and a turning point of type 2 appear. Points of types 4 and 5 (where the MFCQ is violated) cannot appear. Then, we can apply the algorithm JUMP I using the starting point $(x^0, 0)$. $\qquad\qquad\square$

The special embeddings $P^i(t)$, $i\in\{1,2,3\}$, are suitable. $P^1(t)$ requires knowledge of a feasible starting point. This is a restrictive requirement on the one hand, but on the other $M(t) \equiv M$ for all $t\in[0,1]$ and (B6) is a natural assumption.

REMARK 6.2.3 The most restrictive assumption of Theorem 6.2.3 is (B6) (cf. Remark 6.2.1). This assumption may be violated if we consider the embeddings $P^2(t)$.

Consequently we drop (B6) and apply the algorithm JUMP II (cf. Section 5.3).

We assume that the conditions (B1), (B2), (B3′) and (B4) are satisfied. Then we can apply the algorithm JUMP II to find a numerical description of as many connected components in Σ_{gc} as possible. Of course, if we achieve $t = 1$, this algorithm has been convergent. In the other case, the algorithm is not successful. In the worst case we need a numerical description of all connected components in Σ_{gc}. The algorithm JUMP II cannot give the guarantee, since we have jumps only in the situation i, $i \in \{1, 2, 3, 4, 5\}$ (cf. Section 5.3).

6.2.1 Special embeddings

First, we consider the case that a point $x^0 \in M$ is known. Then, we propose the following one-parametric optimization problem as a concretization of (6.2.3):

$$P^1(t): \qquad \min\{f^1(x,t) \mid x \in M\}, \qquad t \in [0,1]$$

where

$$
\begin{aligned}
f^1(x,t) &:= tf(x) + (1-t)\|x - x^0\|^2, \\
h_i^1(x,t) &:= h_i(x), && i \in I, \\
g_j^1(x,t) &:= g_j(x), && j \in J.
\end{aligned}
$$

We see that the properties (B1) and (B2) are satisfied. Moreover, x^0 is even a global minimizer for $P(0)$.

Now, we assume (B4) and that M is non-empty, compact and the MFCQ is satisfied for all $x \in M$.

The assumptions with respect to M are natural ones. (B4) is a generic condition. Using Theorem 6.2.3 the algorithm JUMP I generates an approximation of a local minimizer for the problem (P).

We do not need jumps in the convex case and we find a local minimizer by using the algorithm PATH I (cf. Theorem 6.2.1) and the algorithm PATH II, respectively. Of course, we can weaken the assumption (B4) by (B4').

Secondly, we consider some embeddings for the case in which a feasible point is not known. We assume that $I = \varnothing$. It is easy to see that we can restrict our investigation, by introducing an additional variable as well as one constraint, to problems of the following type:

$$(\mathrm{P}'): \qquad \min\{x_n \mid x \in M\},$$

where

$$M := \{x \in \mathbb{R}^n \mid g_j(x) \leqslant 0, j \in J\}.$$

Now we consider

$$P^2(t): \qquad \min\{f^2(x,t) \mid x \in M^2(t)\}, \qquad t \in [0,1],$$

where

$$M^2(t) := \{x \in \mathbb{R}^n \mid g_j^2(x,t) \leqslant 0, j \in J\},$$

$$f^2(x, t) := tx_n + (1 - t)\|x - x^0\|^2,$$

$$g_j^2(x, t) := g_j(x) + (t - 1)|g_j(x^0)|, \qquad j \in J,$$

and x^0 is arbitrarily chosen.

We denote by $\psi^2(t)$ the set of all global minimizers for $P^2(t)$.

LEMMA 6.2.1 Assume that M is non-empty and g_j, $j \in J$, are continuous functions. Then $P^2(t)$ has the following properties:

(a) (i) $P^2(1) \equiv (P')$,

  (ii) $M^2(t_1) \supset M^2(t_2)$ for $t_1 < t_2$, $t_1, t_2 \in [0, 1]$,

  (iii) $\psi^2(t) \neq \varnothing$ for all $t \in [0, 1)$,

  (iv) $\psi^2(0) = \{x^0\}$.

(b) Assume additionally that the functions $g_j(x) \in C^2(\mathbb{R}^n, \mathbb{R})$ are convex and $\psi^2(1)$ is non-empty and compact. Then we have:

  (i) $\psi^2|[0, 1]$ is u.s.c.-B on $[0, 1]$.

  (ii) $\psi^2(t) = \{x(t)\}$ for all $t \in [0, 1)$ and $x(t)|[0, 1)$ is continuous on $[0, 1)$. Furthermore, each cluster point of any sequence $x(t_k)$ with t_k tending to 1 is contained in $\psi^2(1)$.

  (iii) (V3) is satisfied for all $x \in \psi^2(t)$ for all $t \in [0, 1)$.

Proof (a) (i), (ii), (iv) are obvious.

  (iii) Let $t \in [0, 1)$ be arbitrarily fixed and $x^1 \in M^2(t)$ arbitrary. Using (ii) and that M is assumed to be non-empty, we have $\varnothing \neq M^2(1) \subset M^2(t)$. Furthermore, the function $f^2(x, t)$ is quadratic and strictly convex for $t \in [0, 1)$. Therefore, the existence of global minimizers is guaranteed by standard arguments.

  (b) (i) Let $\tilde{t} \in [0, 1]$, $\tilde{x} \in M^2(\tilde{t})$. It is known that $\varphi(t) := \min\{\|x - \tilde{x}\| \mid x \in M(t)\}$ is a convex function defined on $[0, 1]$. Then we have

$$0 \leqslant \overline{\lim_{t \to \tilde{t}}} \varphi(t) \leqslant \varphi(\tilde{t}) = 0.$$

This means that $\lim_{t \to \tilde{t}} \rho(\tilde{x}, M^2(t)) = 0$ (where $\rho(x, A)$ denotes the distance between the point x and the set A), which shows that $M(t)$ is l.s.c.-B on $[0, 1]$. Then Theorem 4.3.3 [12] implies that $\psi^2(t)$ is u.s.c.-B on $[0, 1]$.

  (ii), (iii) $f(x, t)$ is strictly convex for all $t \in [0, 1)$. Then (ii), (iii) are obvious.

$\square$

For non-convex problems, the embedding $P^2(t)$ will be successful if (B3′), (B4) and (B6) are satisfied (cf. Theorem 6.2.3). If (B6) is violated (cf. Remarks 6.2.1

and 6.2.4), we can try to apply the algorithm JUMP II, but without any guarantee to find a g.c. point for $P^2(1)$.

For convex problems, we can try to apply the algorithm PATH I, but the assumptions (V2) may be violated. Hence, we consider a modified embedding for which (V2) is satisfied. We consider

$$P^3(t): \qquad \min\{f^3(x,t)\,|\,M^3(t)\}, \qquad t\in[0,1],$$

where

$$M^3(t):= \{x\in\mathbb{R}^n\,|\,g_j(x,t)\leqslant 0, j\in J\},$$
$$f^3(x,t):= f^2(x,t) + \|v\|^2,$$
$$g_j^3(x,t):= g_j^2(x,t) + (1-t)v_j, \qquad j\in J$$

(f^2 and g_j^2 are defined in $P^2(t)$).

We denote the set of all global minimizers for $P^3(t)$ by $\psi^3(t)$. Then, by using the same arguments as for the proof of Lemma 6.2.1 we have

LEMMA 6.2.2   Assume that M is non-empty and $g_j\in C^1(\mathbb{R}^n, \mathbb{R})$. Then $P^3(t)$ has the following properties:

(a) (i) $\psi^3(0) = \{(0,0)\}$.

    (ii) $\psi^3(1) = \psi^1(1) \times \{0\}$ if $\psi^1(1) \neq \varnothing$.

    (iii) The LICQ is fulfilled for all $x\in M^3(t)$, $t\in[0,1)$, if the LICQ is satisfied for all $x\in M$.

    (iv) $\psi^3(t) \neq \varnothing$ for all $t\in[0,1)$.

(b) Assume additionally that the functions $g_j\in C^2(\mathbb{R}^n, \mathbb{R})$ are convex and $\psi^1(1)$ is non-empty and compact. Then we have:

    (i) $\psi^3|[0,1]$ is u.s.c.-B on $[0,1]$.

    (ii) $\psi^3(t) = \{(x(t), v(t)\}$ for all $t\in[0,1)$ and $(x(t), v(t))|[0,1)$ is continuous on $[0,1)$. Furthermore, each cluster point of any sequence $x(t_k)$ with t_k tending to 1 is contained in $\psi^3(1)$.

    (iii) (V3) is fulfilled for all $(x,v)\in\psi^3(t)$, $t\in[0,1)$.

Lemma 6.2.2 states that for convex problems the assumptions (E1), (V2) and (V3) restricted to $[0,1)$ and (B1) and (B2) are fulfilled. Furthermore, (b) (ii) allows us to find a local minimizer for $t=1$ by using the algorithm PATH I (cf. Theorem 6.2.1 and the property (b) (ii)). In the non-convex case. we have a quite different situation.

Of course, we can try to apply the algorithm JUMP I under the additional assumption that the LICQ is satisfied for all $x\in M$ (cf. (a) (iii)) and M is a compact set. Then, another difficulty could appear. Since $M^3(t)$ is not compact for $t \in [0, 1)$, $\|v(t)\|$, where $(x(t), v(t))$ are local minimizers for $P^3(t)$, could tend to $+\infty$ for $t \to 1$. Therefore, the sequence generated by the algorithm JUMP I does not necessarily converge to a stationary point of (P) (cf. Figure 6.2 for the only constraint $g_1(x, v, t) \leq 0$). It will even be the exception that $v(t)$ converges to any $\bar{v}$ if t tends to 1. This statement will become clearer if we show the relation to penalty methods in Remark 6.2.4.

REMARK 6.2.4 (On the relation to penalty methods.) The considerations in this remark and Remark 6.2.5 are motivated by personal conversation with J. Stoer and L. Grippo.

To show this relation we discuss the KKT system for $P^3(t)$ for the active index set J_0 (the restriction to the active index set is possible since we follow the active index set strategy in PATH I, II, III):

$$\begin{pmatrix} D_x f^2(x, t) \\ 2v^{\mathrm{T}} \end{pmatrix} + \sum_{j \in J_0} \mu_j \begin{pmatrix} D_x g_j^2(x, t) \\ (1 - t)(e^j)^{\mathrm{T}} \end{pmatrix} = 0 \tag{6.2.5}$$

$$g_j^2(x, t) + (1 - t)v_j = 0, \qquad j \in J_0. \tag{6.2.6}$$

We eliminate v_j, $j \in J_0$, from the system (6.2.6) for $t \in [0, 1)$. Inserting v_j into the objective provides the new unconstrained problem

$$\min \left\{ f^2(x, t) + \frac{\sum_{j \in J_0} (g_j^2(x, t))^2}{(1 - t)^2} \,\middle|\, x \in \mathbb{R}^n \right\} \tag{6.2.7}$$

and

$$\min \left\{ t x_n + (1 - t)\|x - x^0\|^2 + \left(\frac{1}{1 - t} \right)^2 \sum_{j \in J_0} [g_j(x) + (1 - t)|g_j(x^0)|]^2 \,\middle|\, x \in \mathbb{R}^n \right\} \tag{6.2.8}$$

respectively, which are typical of penalty methods. We observe that the expression $(1 - t)\|x - x^0\|^2$ can be interpreted as a regularization term.

The KKT system corresponding to (6.2.7) is

$$D_x f^2(x, t) + \frac{2}{(1 - t)^2} \sum_{j \in J_0} g_j^2(x, t) D_x g_j^2(x, t) = 0. \tag{6.2.9}$$

Then, (6.2.9) is equivalent to the KKT system (6.2.5) and (6.2.6) of $P^3(t)$ in the following sense: if $(x(t), v(t), t)$ is a solution of (6.2.5) and (6.2.6), then $(x(t), t)$ is a solution of (6.2.9), and if $(x(t), t)$ is a solution of (6.2.9), then there exists a $v(t)$ such that $(x(t), v(t), t)$ is a solution of (6.2.5) and (6.2.6).

If we do not follow the active index set strategy, an interpretation as a penalty method is possible, too.

The auxiliary variables v_j can be directly computed for $t \in [0, 1)$ as follows:

$$v_j = -\frac{1}{1-t} \max\{g_j^2(x), 0\}, \qquad j \in J.$$

Then we obtain

$$\min\left\{ tx_n + (1-t)\|x-x^0\|^2 + \left(\frac{1}{1-t}\right)^2 \sum_{j \in J} [\max\{g_j(x)+(1-t)|g_j(x^0)|, 0\}]^2 \,\Big|\, x \in \mathbb{R}^n \right\}.$$

$$(6.2.10)$$

This consideration implies that the numerical effort does not increase by using $P^3(t)$ instead of $P^2(t)$. Moreover, the sequence $(x(t^k), t^k)$ obtained by pathfollowing methods for the problem $P^3(t)$ or generated by penalty methods will be the same if we apply any locally convergent algorithm in each step by choosing the right step size $t_{i+1} - t_i$.

This leads to the following first conclusions:

(i) The LICQ is satisfied automatically at each generated point $(x(t), v(t)) \in M^3(t)$.

(ii) For convex problems, the rate of convergence of the penalty method is at least superlinear by using locally convergent algorithms.

(iii) We have the following observations for non-convex problems:

(a) As noted above,

$$v_j(t) = -\frac{1}{1-t} g_j^2(x(t), t), \qquad j \in J_0,$$

may tend to $-\infty$ for $t \to 1$ if we numerically follow the connected component $\{(x(t), v(t), t)|t \in [t_A, 1)\}$ (where $t_A \geqslant 0$) in the set of local minimizers for $P^3(t)$. In this case we need a jump to another connected component, but this is an open question.

(b) Turning points of type 2 and type 3 could occur. This means that the path stops in the set of local minimizers at a point $(\bar{x}, \bar{t})$. Then, by increasing the penalty parameter

$$\frac{1}{1-\tilde{t}}, \qquad \tilde{t} > \bar{t},$$

we cannot expect to find a stationary point for the problem (6.2.8) for $t = \tilde{t}$. From our consideration in Section 5.2 we have learnt that we have to compute a direction of descent in the non-classical way.

(c) To avoid the difficulty described in (a) we can consider an additional constraint like

$$\| (x, v) \|^2 \leqslant q,$$

where $q \in \mathbb{R}$ is sufficiently large. Then, the feasible set is compact for all $t \in [0, 1]$, but the MFCQ could be violated for some t (cf. $P^4(t)$ and Figure 6.2).

(d) From (a) and (c) it follows that we cannot expect a parametrization that fulfils the assumption of Theorem 6.2.3 surely.

The next embedding illustrates Theorem 6.2.2. We consider the problem

(P) $$\min \{ f(x) | h_i(x) = 0, i \in I, g_j(x) \leqslant 0, j \in J \}$$

and assume

(C1) $\varnothing \neq M := \{ x \subset \mathbb{R}^n | h_i(x) = 0, i \in I, g_j(x) \leqslant 0, j \in J \}$ and M is compact,

(C2) for each stationary point $\bar{x}$ of (P) it holds:

(a) the LICQ is satisfied at $\bar{x}$,

(b) $\mu_j > 0$, $j \in J_0(\bar{x})$,

(c) $D^2 L(\bar{x}) | T_{\bar{x}} M$ is non-singular.

Condition (C1) implies that there is a convex three-times continuously differentiable function $g_{s+1} : \mathbb{R}^n \to \mathbb{R}$ such that $x \in C(r)$ for all $r \in [r_1, r_2]$, $r_1 < r_2$, where $C(r) := \{ x \in \mathbb{R}^n | g_{s+1}(x) \leqslant r \}$ with $C(r)$ compact and $\operatorname{int} C(r) \neq \varnothing$ (cf. $g_{s+1}(x) := \| x \|^2$).

Next we assume that (P), extended by the constraint $g_{s+1}(x) \leqslant r$, satisfies the following constraint qualification on $C(r)$:

(C3) For each $r \in [r_1, r_2]$ and for each $x \in C(r)$ it holds:

(a) the set $\{ D_x h_i(x) | i \in I \}$ is linearly independent,

(b) there exists a vector $\xi \in \mathbb{R}^n$ with

$$h_i(x) + D_x h_i(x) \xi = 0, \qquad i \in I,$$
$$g_j(x) + D_x g_j(x) \xi < 0, \qquad j \in J, \text{ with } g_j(x) \geqslant 0,$$
$$d_x g_{s+1}(x) \xi < 0 \qquad \text{if } g_{s+1}(x) = r.$$

(C3) is a certain extension of the MFCQ to a 'suitable' neighbourhood of M. The meaning of suitable will become clear if we consider the concrete embedding $P^4(t)$.

Finally, we assume

(C4) $f, h_i, g_j, i \in I, j \in J$, are three-times continuously differentiable.

Under the assumptions (C1)–(C4) we consider the following parametric optimization problem

$$P^4(t): \qquad \min\{f(y,t)\,|\,h_i(y,t)=0, i \in I, g_j(y,t) \leqslant 0, j \in \bar{J}\},$$

where

$$\bar{J} = J \cup \{s+1, s+2\}, \qquad y = (x,v) \in \mathbb{R}^n \times \mathbb{R}^{m+s}$$
$$f(y,t) := t[f(x) + \tfrac{1}{2}\|v\|^2] + (1-t)\tfrac{1}{2}(\|x - x^0\|^2 + \|v - v^0\|^2)$$
$$h_i(y,t) := th_i(x) + (1-t)(v_j - \alpha_j), \qquad j \in I,$$
$$g_j(y,t) := tg_j(x) + (1-t)(v_j - \alpha_j), \qquad j \in J,$$
$$g_{s+1}(y,t) := g_{s+1}(x) - r,$$
$$g_{s+2}(y,t) := \|v\|^2 - q,$$

for a fixed vector $w = (x^0, v^0, d, r, q) \in \mathbb{R}^n \times \mathbb{R}^{m+s} \times \mathbb{R}^{m+s} \times [r_1, r_2] \times (0, +\infty)$.

It is easy to see that any stationary point y^* of $P^4(1)$ is of the form $y^* = (x^*, 0)$ and corresponds to a stationary point x^* of (P), and vice versa.

We denote by $M^4(y,t)$ the feasible set of $P^4(y,t)$ and by $\mathcal{H}^4(y, \lambda, \mu, t)$ the mapping corresponding to $P^4(t)$ and defined by (4.1.1).

LEMMA 6.2.3   Assume that (P) satisfies (C1)–(C4) and let

$$W := \{w = (x^0, v^0, \alpha, r, q) \in \mathbb{R}^n \times \mathbb{R}^{m+s} \times \mathbb{R}^{m+s} \times \mathbb{R} \times \mathbb{R} \,|\, g_{s+1}(x^0) < r, \|d\| < q,$$
$$v_j^0 > d_j > 0, j \in J, r_1 < r < r_2, q > 0\}.$$

Then, for almost all $w \in W$, the point $(y^0, \lambda^0, \mu^0, 0)$ with $y^0 = (x^0, d)$, $(\lambda^0, \mu^0) = (v^0 - d, 0)$ is the only KKT point for $P^4(0)$, and the connected component S_0 in the set of KKT points containing $(x^0, d, \lambda^0, \mu^0, 0)$ is a PC^1 path that connects $(x^0, d, \lambda^0, \mu^0, 0)$ with some point $(x^*, 0, \lambda^* \mu^*, 1)$, where x^* is a stationary point for (P).

Proof   It is easy to see that (y^0, λ^0, μ^0) is the only KKT point for $P^4(0)$ and that $P^4(0)$ is strongly regular in the sense of Robinson (cf. Theorem 4.1 in [189]). Then, by Theorem 2.1 in [189], $P^4(t)$ has a unique solution for each t in some neighbourhood of 0. Now, we show that the MFCQ is satisfied for all $y \in M^4(0)$:

Let $y = (x, v) \in M^4(0)$. Then, $D_y h_i(y, 0) = (e_{n+i})^T, i \in I, D_y g_j(y, 0) = (e^{n+j})^T, j \in J$, where e^{n+k} is the $(n+k)$th natural basis vector in $\mathbb{R}^n \times \mathbb{R}^m \times \mathbb{R}^s$. Therefore, $\{D_y h_i(y, 0), i \in I\}$ is linearly independent, and we can see that $\xi \in \mathbb{R}^n \times \mathbb{R}^m \times \mathbb{R}^s$ with

$$\xi_\alpha := x_\alpha^0 - x_\alpha, \qquad \alpha \in \{1, \ldots, n\},$$
$$\xi_i := 0, \qquad i \in I,$$
$$\xi_j := -v_j, \qquad j \in J,$$

is an MF vector.

Moreover, we show that the MFCQ is satisfied for all $y \in M^4(t)$, $t \in [0, 1)$, too. Let $y = (x, v) \in M^4(t)$. Then we have

$$D_y h_i(y, t) = (t D_x h_i(x), (1 - t)(e_{n+i})^{\mathrm{T}}), \qquad i \in I,$$
$$D_y g_j(y, t) = (t D_x g_j(x), (1 - t)(e_{n+j})^{\mathrm{T}}), \qquad j \in J.$$

Then, $\{D_y h_i(y, t), i \in I\}$ is linearly independent. Now, we consider the index set J and we introduce the following subset

$$J^- = \{J \in J_0(y, t) | g_j(x) < 0\}.$$

For $j \in J^-$ by using (C3)(b) there is a number $\lambda_j > 1$ with $\lambda_j g_j(x) + D_x g_j(x) < 0$ and

$$v_j - d_j = -\frac{t}{1-t} g_j(x) > 0,$$

hence $v_j > d_j > 0$ Then, we can see that $\xi \in \mathbb{R}^n \times \mathbb{R}^m \times \mathbb{R}^s$, given by

$$\xi_\alpha = x_\alpha, \qquad \alpha \in \{1, \ldots, n\},$$
$$\xi_i = d_i - v_i, \qquad i \in I,$$
$$\xi_j = d_j - v_j, \qquad j \in J \setminus J^-,$$
$$\xi_j = \lambda_j(d_j - v_j), \qquad j \in J^-,$$

is an MF vector.

(C3) implies directly that the MFCQ is fulfilled for all $y \in M^4(1)$. Of course, $M^4(t) \subset \bar{C}(r, q) := C(r) \times \{v \in \mathbb{R}^n | \|v\|^2 \leq q\}$ and $\bar{C}(r, q)$ is compact. It remains to show that 0 is a regular value of $\mathcal{H}^4$ for almost all $w \in W$ (by using the special structure of this mapping and the parametrized Sard's lemma ((B2) guarantees full rank at $t = 1$), cf. Kojima [134], remark after Theorem 7.1). Therefore, by applying the main statement in the proof of Theorem 6.2.2 the theorem is proved. $\qquad \qquad \square$

For more details of the proof we refer to Theorem 4.2 in Gfrerer *et al.* [71].

We observe that assumption (C3) implies that the MFCQ is satisfied for $y \in M^4(t)$, $t \in [0, 1]$. As noted in Remark 6.2.1, $M^4(0)$ is homeomorphic to M. Therefore, (C3) restricts the class of non-convex problems. In general, this assumption is not satisfied (cf. Figure 6.1). This is the prize for the compactification of $M^4(t)$, $t \in [0, 1)$ (cf. Remark 6.2.5 (iii)(c)).

Of course, if the MFCQ is violated in some stationary points and (B4) is satisfied, the path of stationary points stops at such points (type 4 or type 5 (where the MFCQ is not fulfilled)) and we can try to realize the algorithm JUMP II in the set of g.c. critical points.

REMARK 6.2.5 (On the relation to penalty methods, continued.) For simplicity we assume that $x^0 = v^0 = d^0 = 0$. Furthermore, the last two constraints are ignored in the first consideration. Then, the auxiliary variables v_j can be directly computed for $t \in [0, 1)$:

$$v_i = -\frac{t}{1-t} h_i(x), \qquad i \in I,$$

$$v_j = -\frac{t}{1-t} \max\{g_j(x), 0\}, \qquad j \in J.$$

Then, we obtain (cf. 6.2.10)

$$\min\left\{ tf(x) + \tfrac{1}{2}(1-t)\|x\|^2 + \tfrac{1}{2}\left(\frac{t}{1-t}\right)^2 \left(\sum_{i \in I} h_i(x)^2 + \sum_{j \in J} [\max\{g_j(x), 0\}]^2 \right) \Big| x \in \mathbb{R}^n \right\}.$$

$$(6.2.11)$$

Of course, if we ignore the latter two constraints, we have the same difficulties as described in Remark 6.2.4 (iii)(a). Hence, we ask the question: What is the translation of the 'full' problem $P^4(t)$ to a penalty method?

We denote the objective function of (6.2.11) by $p(x, t)$ (see Figure 6.2). Then the problem corresponding to $P^4(t)$ is the following:

$$\min\left\{ p(x, t) \Big| g_{s+1}(x) \leq r, \left(\frac{t}{1-t}\right)^2 \left(\sum_{i \in I} h_i(x)^2 + \sum_{j \in J} [\max\{g_j(x), 0\}]^2 \right) \leq q \right\},$$

where $q > 0$ is sufficiently large.

We note that the objective in (6.2.11) is a non-differentiable function. Using the active index set strategy we obtain a sequence of differentiable problems.

The most important constraint for our consideration is the second one. This means that the penalty term including the penalty parameter has an upper bound for increasing penalty parameters. In the case when the upper bound is achieved, the MFCQ is violated and one of the difficult singularities (type 4 or type 5 (MFCQ is not fulfilled)) will appear (see Figure 6.2).

6.3 ON GLOBAL OPTIMIZATION

We consider the problem of global optimization (cf. (1.2.5))

$$\text{glob} \min\{F(x) | x \in \mathbb{R}^n\}, \qquad (6.3.1)$$

i.e. find a global minimizer of $F(x)$ on $\mathbb{R}^n$.

We follow here J. Guddat and H. Th. Jongen [85] and:

(D1) $F \in C^{\infty}(\mathbb{R}^n, \mathbb{R})$,

(D2) F has compact lower level sets, i.e. $\{x \in \mathbb{R}^n \,|\, F(x) \leqslant \alpha\}$ is compact for all $\alpha \in \mathbb{R}$.

Of course, (D2) restricts the class of functions considered. However, if we follow the approach proposed in Section 1.2, we have only one constraint in problem (6.2.3). Then the analysis will be simplified. However, we follow the same strategy as proposed in Section 1.2.

Step 1: Compute a stationary point $\hat{x}$ for

$$\min\{F(x) \,|\, x \in \mathbb{R}^n\}.$$

Step 2: Find a point belonging to

$$E(\varepsilon) := \{x \in \mathbb{R}^n \,|\, F(x) \leqslant F(\hat{x}) - \varepsilon\} \tag{6.3.2}$$

with $\varepsilon > 0$ sufficiently small.

As already noted in Section 1.2, step 2 is the difficult one and the subject of our discussion in this section. We note that the set $E(\varepsilon)$ is compact, since (D2) is satisfied.

Now we consider an arbitrary point $x^0 \subset \mathbb{R}^n$ and an arbitrarily chosen function $f \in C^{\infty}(\mathbb{R}^n, \mathbb{R})$ with the following property:

(D3) x^0 is a global minimizer for $\min\{f(x) \,|\, x \in \mathbb{R}^n\}$ and
$\quad\;\; x^0$ is the only stationary point for $\min\{f(x) \,|\, x \in \mathbb{R}^n\}$.

In Section 1.2 such an f was defined by $f(x) = \|x - x^0\|^2$.

Of course, step 2 can be realized if we can compute a g.c. point of the following problem

$(\hat{P})$:
$$\min\{f(x) \,|\, g(x) \leqslant 0\}, \tag{6.3.3}$$

where $g(x)$ is defined by

$$g(x) := F(x) - F(\hat{x}) + \varepsilon. \tag{6.3.4}$$

As in Section 1.2 we follow the concept of embedding and propose the following one-parametric optimization problem

$\hat{P}(t)$:
$$\min\{f(x, t) \,|\, g(x, t) \leqslant 0\}, \qquad t \in [0, 1], \tag{6.3.5}$$

where

$$f(x, t) := f(x), \tag{6.3.6}$$

$$g(x, t) := g(x) + (t - 1)g(x^0), \tag{6.3.7}$$

and

(D4) $g(x^0) > 0$.

Of course, unless (D4) is satisfied, we have $F(x^0) \leqslant F(\hat{x}) - \varepsilon$ and setp 2 is realized.

We recall that Diener [37] discussed the problem of global optimization also from the point of view of homotopy by connecting all critical points of a smooth function.

We see that $\hat{P}(t)$ is a special case of the problem $P(t)$ introduced e.g. in (5.1.1). We denote the feasible set of $(\hat{P})$ and $\hat{P}(t)$ by $\hat{M}$ and $\hat{M}(t)$, respectively, and the set of all global (local) minimizers for $\hat{P}(t)$ by $\hat{\psi}(t)$ ($\hat{\psi}_{\text{loc}}(t)$).

Then $\hat{x}$ solves the problem (6.3.1) if and only if $E(\varepsilon)$ defined by (6.3.2) is empty for $\varepsilon > 0$ sufficiently small. Here, we do not discuss the question: how can it be checked whether $E(\varepsilon)$ is empty or not? This is the reason why we assume:

(D5) $\hat{M} \neq \varnothing$.

We observe that $\hat{P}(t)$ has the following properties:

(F1) x^0 is a global minimizer for $\hat{P}(0)$,

(F2) $\hat{P}(1) \equiv (\hat{P})$,

(F3) $\hat{M}(t_1) \supset \hat{M}(t_2)$ for $t_1 \leqslant t_2$, $t_1, t_2 \in [0, 1]$ (cf. (D4)),

(F4) $\hat{M}(t)$ is non-empty for all $t \in (-\infty, 1]$ (cf. (D5) and (F3)) and there exists a compact set K with $\hat{M}(t) \subseteq K$ for all $t \in (-\infty, 1]$ (cf. (D2)),

(F5) $\hat{\psi}(t)$ is non-empty for all $t \in (-\infty, 1]$ (cf. (F4)).

Now we assume for the problem $\hat{P}(t)$ defined by (6.3.5), (6.3.6) and (6.3.7):

(D6) $(f, g) \in \mathscr{F}^{**}$.

Furthermore, we use the same notations for $\hat{P}(t)$ as for the general problem $P(t)$ like Σ_{gc}, Σ_{gc}^i, $i \in \{1, \dots, 5\}$, and Σ_{loc}. First, we ask the question: what kind of singularities can appear? The answer will be given by the following

THEOREM 6.3.1   Let (D1)–(D6) be fulfilled. Then we have the following:

(i) If $\bar{z} = (\bar{x}, \bar{t}) \in \bar{\Sigma}_{\text{loc}} \setminus \Sigma_{\text{gc}}^1$ with $\bar{t} \in (0, 1]$, then $\bar{z} \in \Sigma_{\text{gc}}^3 \cup \Sigma_{\text{gc}}^4$.

(ii) $(x^0, 0) \in \Sigma_{\text{loc}} \cap \Sigma_{\text{gc}}^2$ and $(x^0, 0)$ is no turning point.

(iii) If $\bar{z} \in \bar{\Sigma}_{\text{loc}} \cap \Sigma_{\text{gc}}^4$, then there exists a local C^∞ coordinate transformation (5.2.2) sending $\bar{z}$ onto the origin, such that g takes the form

$$g(x, t) = \sum_{i=1}^{n} x_i^2 + t$$

(in these new coordinates).

Proof   (i) First, we consider $\bar{z} \in \bar{\Sigma}_{\text{loc}}$ and suppose that $\bar{z} \in \Sigma_{\text{gc}}^i$, $i \in \{2, \dots, 5\}$. Let $\bar{z} \in \Sigma_{\text{gc}}^2$. We know that x^0 is a global minimizer and the only stationary point for $\min\{f(x)\,|\,x\in\mathbb{R}^n\}$ (cf. (D3)) and that

$$g(x^0, 0) = 0 \tag{6.3.8}$$

and $g(x^0, t) > 0$ for all $t\in(0, 1]$ (cf. (6.3.6) and (D4)). As we have only one constraint, $\bar{z}\in\Sigma^2_{\mathrm{gc}}$ does not exist for $t\in(0, 1]$. Let $\bar{z}\in\Sigma^5_{\mathrm{gc}}$. Then the property (D1) in (2.5.25) cannot be fulfilled for an arbitrary $n\geqslant 1$. Thus, points of type 5 cannot occur.

(ii) By application of (6.3.7) and (D4) it is easy to see that

$$g(x^0, t) < 0 \qquad \text{for } t < 0. \tag{6.3.9}$$

Now x^0 is a global minimizer for $\hat{P}(t)$, $t\leqslant 0$. Using (6.3.8) and (6.3.9) we obtain (ii).

(iii) From Section 5.2 we know that there exists a local C^∞ coordinate transformation sending $\bar{z}$ onto the origin which can be constructed such that g takes the following form:

$$g(x, t) = -\sum_{i=1}^{k} x_i^2 + \sum_{j=k+1}^{n} x_j^2 - \delta t$$

in these new coordinates, where $\delta\in\{+1, -1\}$.

Using (6.3.7) we obtain $\delta = \mathrm{sign}(-D_t g(\bar{x}, \bar{t})) - \mathrm{sign}(-g(x^0))$. As $g(x^0) > 0$ holds (cf. (D4)), we have $\delta = -1$ and (iii) is proved. $\qquad\square$

REMARK 6.3.1   If $\bar{z}\in\bar{\Sigma}_{\mathrm{loc}}$ is a point of type 4, then $\bar{z}$ fulfils condition (5.2.32). We remind the reader of the fact that this case was the difficult one; namely, if f increases, the corresponding connected component of the feasible set $\hat{M}(t)$ will shrink to one point and become empty for increasing t (cf. Figure 5.10, IIb). Figure 6.1 shows that this situation is typical of the application of the embedding $\hat{P}(t)$ defined by (6.3.5), (6.3.6) and (6.3.7) to the problem of global optimization.

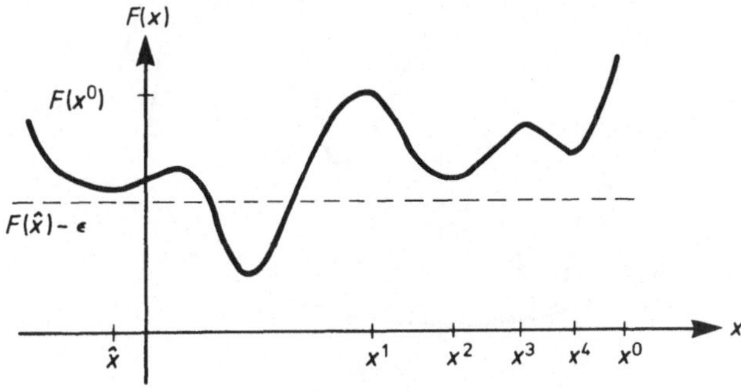

Figure 6.1

Now $\hat{x}$ is the local minimizer computed by step 1, and x^0 is the starting point for step 2. Then we see that the MFCQ is not fulfilled at the parameter values $t \in [0, 1]$, where $g(x^i, t) = 0$, $i = 1, 2, 3, 4$, and $g(\hat{x}, t) = 0$. As noted in Section 6.1, we still do not have a proposal working for a possible jump in $\bar{\Sigma}_{\text{loc}}$. Therefore, the algorithm JUMP I (for increasing t, find a discretization (5.1.4) and corresponding local minimizers $x(t_i)$, $i = 1, \ldots, N$) is not successful. Consequently, we propose to apply the algorithm JUMP II to follow numerically the connected component $C = C(x^0, 0)$ in Σ_{gc}. Jumps are possible in the situation i, $i \in \{1, 2, 4\}$. The situation j, $j \in \{3, 5\}$, cannot appear.

REMARK 6.3.2   (i) The analysis included in this section allows a deeper insight into the essential difficulties of finding a successful deterministic algorithm for the problem of global optimization.

Of course, if we can find a jump at a point of type 4 with (5.2.2), a foundation for solving this problem is given for the class $\mathscr{F}^{**}$.

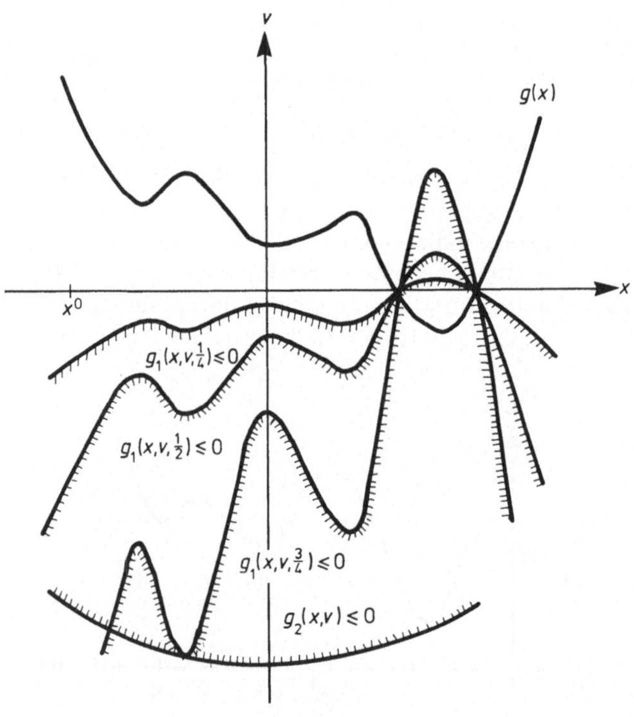

Figure 6.2

(ii) Since we do not have a proposal for a possible jump at a point of type 4 with (5.2.32), we apply the algorithm JUMP II maybe with different starting points $x^0 \in \mathbb{R}^n$ if we are not successful. This can be helpful, in particular in combination with other techniques, e.g. stochastic research procedures. However, there are examples where the chosen starting point x^0 does not lie in a radius of convergence (for a given locally convergent algorithm) for any g.c. point of $\hat{P}(1)$, but $(x^0, 0)$ lies on a curve of g.c. points leading to a g.c. point for $P(1)$. Thus, it is an advantage without doubt.

As noted before, the assumption (D2) is a restrictive one. However, there is a similar analysis if we consider the problem glob $\min\{F(x)|x \in K\}$ (cf. 1.2.2) where K is a given compact subset of $\mathbb{R}^n$, e.g. $K = K(q_0) := \{x \in \mathbb{R}^n \mid \|x\|^2 \leqslant q_0\}$, $q_0 > 0$, and the corresponding parametric optimization problem

$$\hat{P}(t): \qquad \min\{f(x)|x \in \hat{M}(t)\}, \qquad t \in [0, 1], \tag{6.3.10}$$

where (cf. 6.3.5)

$$\hat{M}(t) = \{x \in K \mid g(x, t) \leqslant 0\}. \tag{6.3.11}$$

Of course, (D2) can be omitted in this case. If we choose $K = \{x \in \mathbb{R}^n \mid \|x\|^2 \leqslant q_0\}$, then we have two constraints in $\hat{P}(t)$ defined by (6.5.10).

Now we consider another embedding proposed in Section 6.2 (cf. $P^4(t)$) with regard to the application in global optimization, namely

$$\bar{P}(t): \qquad \min\left\{\left\|\binom{x}{v} - \binom{x^0}{v^0}\right\|^2 \middle| \binom{x}{v} \in \bar{M}(t)\right\}, \qquad t \in [0, 1],$$

where

$$\bar{M}(t) := \left\{\binom{x}{v} \in \mathbb{R}^{n+1} \middle| g_i(x, v, t) \leqslant 0, i = 1, 2\right\},$$

$$g_1(x, v, t) := tg(x) + (1 - t)(v - d),$$

$$g_2(x, v) := \left\|\binom{x}{v}\right\|^2 - q_0,$$

and $q_0 > 0$ sufficiently large and $d \in \mathbb{R}$ fixed.

Figure 6.2 illustrates the typical situation for this embedding (for $d = 0, v^0 = 0$). We see that the properties analogous to (F1), (F2), (F4) and (F5) are fulfilled (of course, we have to choose d in such a way that $d \geqslant v^0$). Furthermore, by choosing (x^0, v^0), q_0 as in Figure 6.2, the MFCQ is not fulfilled for certain parameter values t if we follow the four curves of g.c. points on which (x^0, v^0) lies. We have points of type 4 with (5.2.32) at these parameter values. We cannot jump to another curve of local minimizers, but we reach $t = 1$ by walking on the curve of g.c. points. As we have the special case $n = 1$, this is possible by choosing an arbitrary starting point $(x^0, v^0) \in \mathbb{R}^2$. In the general case $(n \geqslant 2)$ a return to $t = 0$ is possible. From this point of view we have the same situation as in the embeddings proposed before. The only advantage of this parameterization lies in the fact that we have only two g.c. points for $\bar{P}(0)$ (since $\bar{M}(0)$ is a convex polyhedral set). Therefore, we know that we have to jump or to choose another starting point if we return to $t = 0$. We note that the example in Figure 6.2 shows once more that the assumption (B6) in Theorem 6.2.2 and the assumption (C3) in Lemma 6.2.3 are restrictive. For instance, if the function $g(x) \in C^{\infty}(\mathbb{R}^n, \mathbb{R})$ has more than one isolated local minimizer, as in Figure 6.2, these assumptions are not fulfilled for the parametrization $P^4(t)$ considered in Section 6.2.

We can also consider quite another strategy for the problem $\operatorname{glob\,min}\{F(x)\,|\, x \in K\}$ where K is defined e.g. as above

$$K := \{x \in \mathbb{R}^n \,|\, \|x\|^2 \leqslant q_0\}, \qquad q_0 > 0.$$

The problem can be embedded by

$$\tilde{P}(t): \qquad \min\{tF(x) + (1-t)\|x - x^0\|^2 \,|\, x \in K\}, \qquad t \in [t_A, t_B],$$

where $x^0 \in K$ is a starting point and the interval $[t_A, t_B]$ is chosen as in Section 5.3. Then we can apply the algorithm JUMP II to find as many connected components in Σ_{gc} as possible. The advantages lies in the fact that the MFCQ is satisfied for all $x \in K$. Therefore, points of type 4 and type 5 (where the MFCQ is violated) cannot occur. Of course, if we can describe all connected components in Σ_{gc} numerically, all critical points for $\tilde{P}(1)$ will be found and the problem of global optimization will be solved, but we still do not have any estimation (as mentioned in Section 6.1) of the number of connected components in Σ_{gc}. Therefore, we cannot be sure to have found all critical points for $\tilde{P}(1)$.

6.4 ON MULTI-OBJECTIVE OPTIMIZATION

We will discuss here the three different parametrizations $P_i(t)$, $i = 1, 2, 3$, introduced in Section 1.2 (cf equation (1.2.22)) as typical examples in multi-objective optimization. We will compare the use of pathfollowing methods in order to solve $P_i(t)$, $t \in [0, 1]$, $i = 1, 2, 3$, with the traditional approach to solve the optimization problems $P_i(1)$. Furthermore, we will consider these parametrizations with respect to the arising singularities.

The idea of the reference point optimization was developed in several papers (cf. e.g. Wierzbicki [227]) and used e.g. in the program system DIDAS (cf. Grauer *et al.* [76]). This approach consists of trying to solve the problem $P_i(1)$, $i \in \{1, 2, 3\}$, directly. We take quite another approach and propose to use pathfollowing methods with jumps for the one-parametric optimization problem of the type $P_i(t)$, $t \in [0, 1]$, $i \in \{1, 2, 3\}$. We see that a starting point for $P_i(0)$, $i \in \{1, 2, 3\}$, is known or easy to construct.

Now, the main advantages in using pathfollowing methods with jumps instead of solving the problem $P_i(1)$, $i \in \{1, 2, 3\}$, directly will be discussed. For this purpose we make some preconsiderations. We note that the decision-maker is mainly interested in getting to know whether his wish expressed by μ^1 (cf. p. 17) was realistic or not. Moreover, if the goal point μ^1 is a realistic one, then he wants to find a point $\hat{x} \in \mathbb{R}^n$ with

$$\hat{x} \in M, \qquad f_j(\hat{x}) \leqslant \mu_j^1, \qquad j = 1, \ldots, L. \tag{6.4.1}$$

This consideration leads to the following

DEFINITION 6.4.1 (i) μ^1 is called a *realistic goal* if

$$\hat{M}(\mu^1) := \{x \in M \mid f_j(x) \leqslant \mu_j^1, j = 1, \ldots, L\}, \tag{6.4.2}$$

is non-empty.

(ii) A point $\hat{x} \in \hat{M}(\mu^1)$ is called a *goal realizer*.

We assume in the following.

(W1) M is non-empty and compact,

(W2) μ^1 is a realistic goal.

(W1) is a natural assumption. (W2) is a technical assumption for further investigation. First, we consider the problems $P_i(1)$, $i = 1, 2, 3$ (traditional approach), and pose the following question.

QUESTION 6.4.1 How useful are the problems $P_i(1)$, $i = 1, 2, 3$, for finding a goal realizer?

The answer is given in the following remark.

REMARK 6.4.1 (i) If (x, v) is a stationary point for $P_1(1)$ or $P_3(1)$, x is not necessarily a goal realizer unless $v \leqslant 0$.

(ii) Even a global minimizer of $P_2(1)$ is not necessarily a goal realizer.

(iii) Given x^0, we know a starting feasible point and can use descent procedures to find a stationary point for $P_i(1)$, $i = 1, 2, 3$, but it is not necessarily a goal realizer (cf. (i) and (ii)).

Secondly, we consider the one-parametric optimization problems $P_i(t)$, $t \in [0, 1]$, $i = 1, 2, 3$, and pose the next question.

QUESTION 6.4.2 What are the advantages when using solution algorithms (pathfollowing methods with jumps) for one-parametric optimization problems?

The answer is given in the following.

REMARK 6.4.2 (i) In each step (to calculate $x(t_{i+1})$ starting at $x(t_i)$) we can use locally convergent algorithms having better rates of convergence and a good behaviour on the computer (cf. Chapter 3 and 4).

(ii) All points $x(t_i)$, $i = 1, \ldots, N$, could be of interest for the decision-maker (in case $x(t_i) \in \psi_j(t)$ $(x(t_i) \in \psi_{j\text{loc}}(t))$, then $x(t_i) \in M_{\text{eff}}$ or M_{weff} (resp. M_{loceff} or M_{locweff})), i.e. we have a reduction of the computation time for the next iteration point that could be used for the dialogue procedure.

(iii) The chance to find a goal realizer is much greater for $P_1(t)$ and $P_3(t)$ by using pathfollowing methods as the following simple example (cf. Figure 6.3) shows: (x^0, v^0) is a starting point and x^0 is a goal realizer for μ^0. Solving the problem $P_1(1)$ we obtain the local minimizer (x^1, v^1) with $f_j(x^1) > \mu_j^1$, $j = 1, 2$, i.e. x^1 is not a goal realizer. However, using the pathfollowing technique we can obtain (x^2, v^2) and x^2 is a goal realizer.

Thirdly, we compare the different parametrizations under the point of view of using pathfollowing methods with jumps. This leads to the following.

QUESTION 6.4.3 What kind of singularities may appear if we assume that the corresponding functions belong to the class $\mathscr{F}^{**}$ (cf. Chapter 2)?

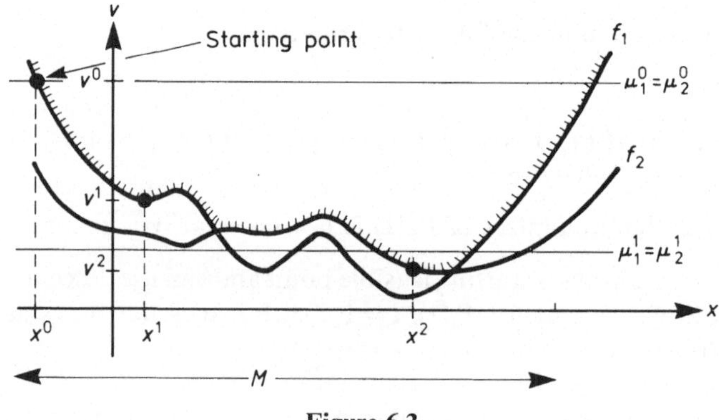

Figure 6.3

Before giving the answer to this question we must specify the five types of generalized critical points for each parametrization.

First parametrization

$$P_1(t): \quad \min\left\{\delta \sum_{i=1}^{L} \lambda_i^0 [f_i(x) - \mu_i(t)] + v \,\Big|\, (x,v) \in M_1(t)\right\}, \qquad t \in [0,1],\ z = (x,t),$$

where

$$M_1(t) = \{(x,v) \in \mathbb{R}^n \times \mathbb{R} \mid x \in M,\ \lambda_i^0 [f_i(x) - \mu_i(t)] \leqslant v,\ i = 1,\ldots,L\},$$
$$\delta > 0, \qquad \lambda_i^0 > 0, \qquad i = 1,\ldots,L,$$
$$\mu_i(t) = \mu_i^0 + t(\mu_i^1 - \mu_i^0), \qquad i = 1,\ldots,L,$$
$$M = \{x \in \mathbb{R}^n \mid h_i(x) = 0,\ i \in I,\ g_j(x) \leqslant 0,\ j \in J\}.$$

A point $(\bar{x}, \bar{v}, \bar{t})$ with $(\bar{x}, \bar{v}) \in M_1(\bar{t})$ is called a g.c. point for $P_1(\bar{t})$ if there exist numbers $\gamma,\ \alpha_i \in I,\ \beta_j,\ j \in J_0(\bar{x}),\ \gamma_k,\ k \in K_0(\bar{x}, \bar{v}, \bar{t})$, not all vanishing, such that

$$\gamma\delta \sum_{i=1}^{L} \mu_i^0 Df_i(\bar{x}) + \sum_{i \in I} \alpha_i Dh_i(\bar{x}) + \sum_{j \in J_0(\bar{x})} \beta_j Dg_j(\bar{x}) + \sum_{k \in K_0(\bar{x}, \bar{v}, \bar{t})} \gamma_k \lambda_k^0 Df_k(\bar{x}) = 0, \quad (6.4.3)$$

$$\gamma - \sum_{k \in K_0(\bar{x}, \bar{v}, \bar{t})} \gamma_k = 0, \qquad\qquad (6.4.4)$$

where

$$J_0(\bar{x}) = \{j \in J \mid g_j(\bar{x}) = 0\}, \qquad K_0(\bar{x}, \bar{v}, \bar{t}) = \{k \in \{1,\ldots,L\} \mid \lambda_k^0 [f_k(\bar{x}) - \mu_k(\bar{t})] = \bar{v}\}.$$

The set $M_1(\bar{t})$ is said to be regular at $(\bar{x}, \bar{v})$ if

$$\left\{\begin{pmatrix} Dh_i(\bar{x}) \\ 0 \end{pmatrix},\ \begin{pmatrix} Dg_j(\bar{x}) \\ 0 \end{pmatrix},\ \begin{pmatrix} Df_k(\bar{x}) \\ -1 \end{pmatrix},\ i \in I,\ j \in J_0(\bar{x}),\ k \in K_0(\bar{x}, \bar{v}, \bar{t})\right\} \qquad (6.4.5)$$

is a linearly independent set.

A g.c. point $(\bar{x}, \bar{v}, \bar{t})$ is called a critical point if $M_1(\bar{t})$ is regular at $(\bar{x}, \bar{v})$.

In analogy to Chapter 2, a critical point $(\bar{x}, \bar{v}, \bar{t})$ is called a point of type 1 if the following conditions hold:

(ND1-1) $\beta_j \neq 0, \qquad j \in J_0(\bar{x}), \qquad \gamma_k \neq 0, \qquad k \in K_0(\bar{x}, \bar{v}, \bar{t}),$

(ND2-1) $D_x^2 L(\bar{x}, \bar{t}) \mid T(\bar{x})$ is non-singular,

where

$$L(x,t) = \delta \sum_{i=1}^{L} \lambda_i^0 [f_i(x) - \mu_i(t)] + \sum_{i \in I} \alpha_i h_i(x) + \sum_{j \in J_0(x)} \beta_j g_j(x)$$
$$+ \sum_{k \in K_0(x,z,t)} \gamma_k \lambda_k^0 [f_k(x) - \mu_k(t)], \qquad\qquad (6.4.6)$$

$$T_{(\bar{x})} = \bigcap_{i \in I} \operatorname{Ker} Dh_i(\bar{x}) \cap \bigcap_{j \in J_0(\bar{x})} \operatorname{Ker} Dg_j(\bar{x}) \cap \bigcap_{k \in K_0(\bar{x}, \bar{v}, \bar{t})} \operatorname{Ker} Df_k(\bar{x}).$$

In order better to understand the condition (ND2-1) it is necessary to observe the following relations.

The Lagrange function of $P_1(t)$ is given by

$$\tilde{L}(x,v,t) = \delta \sum_{i=1}^{L} \lambda_i^0 [f_i(x) - \mu_i(t)] + v + \sum_{i \in I} \alpha_i h_i(x) + \sum_{j \in J_0(x)} \beta_j g_j(x)$$

$$+ \sum_{k \in K_0(x,v,t)} \gamma_k \{\lambda_k^0 [f_k(x) - \mu_k(t)] - v\}.$$

Then we have

$$D_x^2 \tilde{L}(x,v,t) = \begin{bmatrix} D_x^2 L(x,t) & 0_n \\ 0_n^T & 0 \end{bmatrix}, \qquad 0_n \in \mathbb{R}^n, \qquad 0 \in \mathbb{R}. \qquad (6.4.7)$$

The tangent space of $M_1(\bar{t})$ at $(\bar{x}, \bar{v})$ is given by

$$\tilde{T}_{(\bar{x},\bar{v})} = \bigcap_{i \in I} \mathrm{Ker}\begin{pmatrix} Dh_i(\bar{x}) \\ 0 \end{pmatrix} \cap \bigcap_{j \in J_0(\bar{x})} \mathrm{Ker}\begin{pmatrix} Dg_j(\bar{x}) \\ 0 \end{pmatrix} \cap \bigcap_{k \in K_0(\bar{x},\bar{v},\bar{t})} \begin{pmatrix} Df_k(\bar{x}) \\ -1 \end{pmatrix}.$$

Therefore,

$$\tilde{T}_{(\bar{x},\bar{v})} = T_{(\bar{x})} \times \{0\}, \qquad 0 \in \mathbb{R}, \qquad (6.4.8)$$

and (6.4.7) and (6.4.8) imply that $D_{(x,v)}^2 \tilde{L}(\bar{x}, \bar{v}, \bar{t}) | \tilde{T}(\bar{x}, \bar{v})$ is non-singular if and only if $D_x^2 L(\bar{x}, \bar{t}) | T(\bar{x})$ is non-singular.

Following Chapter 2 we give a short characterization of points of the types 2, 3, 4 and 5.

Type 2: Condition (ND1-1) is violated for exactly one multiplier.

Type 3: One eigenvalue of $D_x^2 L(\bar{x}, \bar{t}) | T(\bar{x})$ vanishes.

Type 4: Condition (6.4.5) is violated and

$$|I| + J_0(\bar{x})| + |K_0(\bar{x}, \bar{v}, \bar{t})| < n + 1.$$

Type 5: Condition (6.4.5) is violated and

$$|I| + J_0(\bar{x})| + |K_0(\bar{x}, \bar{v}, \bar{t})| = n + 1.$$

Second parametrization

$$P_2(t): \quad \min \left\{ \frac{1}{\rho} \ln \left[\frac{1}{L} \sum_{i=1}^{L} \left(\lambda_i^0 \frac{\tilde{q}_i - f_i(x)}{\tilde{q}_i - [\mu_i^0 + t(\mu_1^1 - \mu_i^0)]} \right)^\rho \right] \middle| x \in M \right\}, \qquad t \in [0,1],$$

where

$$\tilde{q}_i < \min \{f_i(x) | x \in M\}, \qquad i = 1, \dots, L, \qquad \rho > L \qquad \text{are fixed.}$$

If we denote the objective function of $P_2(t)$ by $f(x,t)$, this parametrization can be considered as a particular case of the general parametric optimization problem $P(t)$ considered in Chapter 2 with $h_i(x, t) \equiv h_i(x)$, $i \in I$, $g_j(x, t) \equiv g_j(x)$, $j \in J$, $t \in [0, 1]$.

Third parametrization

$$P_3(t): \qquad\qquad \min\{v \,|\, (x, v) \in M_3(t)\}, \qquad\qquad t \in [0, 1],$$

where

$$M_3(t) = \{(x, v) \in \mathbb{R}^n \times \mathbb{R} \,|\, h_i(x) = 0, i \in I, g_j(x) \leqslant 0, j \in J, f_k(x) - \mu_k(t) \leqslant v, k = 1, \ldots, L\},$$

$$\mu_k(t) = \mu_k^0 + t(\mu_k^1 - \mu_k^0), \qquad k = 1, \ldots, L.$$

We assume that M is a regular set.

A point $(\bar{x}, \bar{v}, \bar{t})$ with $(\bar{x}, \bar{v}) \subset M_3(\bar{t})$ is called a g.c. point for $P_3(t)$ if there exist numbers γ, $\alpha_i \in I$, β_j, $j \in J_0(\bar{x})$, γ_k, $k \in K_0(\bar{x}, \bar{v}, \bar{t})$, not all vanishing, such that

$$\sum_{i \in I} \alpha_i Dh_i(\bar{x}) + \sum_{j \in J_0(\bar{x})} \beta_j Dg_j(\bar{x}) + \sum_{k \in \tilde{K}_0(\bar{x}, \bar{v}, \bar{t})} \gamma_k Df_k(\bar{x}) = 0, \qquad (6.4.9)$$

$$\gamma - \sum_{k \in \tilde{K}_0(\bar{x}, \bar{v}, \bar{t})} \gamma_k = 0, \qquad (6.4.10)$$

where

$$J_0(x) = \{j \in J \,|\, g_j(\bar{x}) = 0\}, \qquad \tilde{K}_0(\bar{x}, \bar{v}, \bar{t}) = \{k \in \{1, \ldots, L\} \,|\, f_k(\bar{x}) - \mu_k(\bar{t}) = \bar{v}\}.$$

A g.c. point $(\bar{x}, \bar{v}, \bar{t})$ is called a critical point if there exist numbers α_i, $i \in I$, β_j, $j \in J_0(\bar{x})$, γ_k, $k \in \tilde{K}_0(\bar{x}, \bar{v}, \bar{t})$, not all vanishing, such that (6.4.9) and (6.4.10) with $\gamma = 1$ hold.

A critical point $(\bar{x}, \bar{v}, \bar{t})$ is called a point of type 1 if the following conditions hold:

(ND1-3) $\beta_j \neq 0$, $\qquad j \in J_0(\bar{x})$, $\qquad \gamma_k \neq 0$, $\qquad k \in \tilde{K}_0(\bar{x}, \bar{v}, \bar{t})$,

(ND2-3) $D_x^2 L(\bar{x}, \bar{t}) | T(\bar{x})$ is non-singular,

where

$$L(x, t) = \sum_{i \in I} \alpha_i h_i(x) + \sum_{j \in J_0(x)} \beta_j g_j(x) + \sum_{k \in \tilde{K}_0(x, v, t)} \gamma_k [f_k(x) - \mu_k(t)],$$

$$T_{(\bar{x})} = \bigcap_{i \in I} \text{Ker}\, Dh_i(\bar{x}) \cap \bigcap_{j \in J_0(\bar{x})} \text{Ker}\, Dg_j(\bar{x}) \cap \bigcap_{k \in \tilde{K}_0(\bar{x}, \bar{v}, \bar{t})} \text{Ker}\, dF_k(\bar{x}).$$

A moment of reflection shows that similar relations to (6.4.7) and (6.4.8) hold for $P_3(t)$, and this explains the formulation of condition (ND2-3).

Similarly to parametrization $P_1(t)$ we can introduce the four kinds of singularities: types 2, 3, 4 and 5.

The satisfaction of the MFCQ is of great importance in order to answer Question 6.4.3. Namely, if the MFCQ is fulfilled, the most complicated degeneracies cannot appear. Another important question for using pathfollowing techniques was asked by Wierzbicki [227]: whether the parametrization is robustly computable? In this case it is necessary that the parameter-dependent feasible sets $M_i(t)$, $i = 1, 3$, are homeomorphic for all $t \in [0, 1]$, i.e. stable on $[0, 1]$ (cf. [87], Definition 1.1).

THEOREM 6.4.1   (i) Assume that M is non-empty. Then $M_1(t)$ is non-empty for all $t \in [0, 1]$. Further, $\psi_1(t)$ is non-empty for all $t \in [0, 1]$ if M is a compact set and $f_i \in C(\mathbb{R}^n, \mathbb{R})$, $i = 1, \ldots, L$.

(ii) Assume that

(a) M is non-empty,

(b) f_k, h_i, $g_j \in C^2(\mathbb{R}^n, \mathbb{R})$, $k = 1, \ldots, L$, $i \in I$, $j \in J$,

(c) the MFCQ is fulfilled for all $x \in M$.

Then we have:

(1) the MFCQ is fulfilled for all $(x, v) \in M_1(t)$ for all $t \in [0, 1]$,

(2) $M_1(t_1)$ is homeomorphic to $M_1(t_2)$ for all $t_1, t_2 \in [0, 1]$.

Proof   (i) It is easy to see that $M_1(t)$ is non-empty. Now we show that $\psi_1(t)$ is non-empty, too. For an arbitrarily fixed $t \in [0, 1]$ we consider the following optimization problems:

$$\min\left\{\delta \sum_{i=1}^{L} \lambda_i^0 \{f_i(x) - [\mu_i^0 + t(\mu_i^1 - \mu_i^0)]\} \,\middle|\, x \in M\right\}$$

and

$$\min\{v \mid \lambda_i^0 \{f_i(x) - [\mu_i^0 + t(\mu_i^1 - \mu_i^0)]\} \leq v, i = 1, \ldots, L, x \in M\}.$$

Since M is compact, both problems have a global minimizer. Then $\psi_1(t)$ is obviously non-empty.

(ii) (1) Let $\tilde{x} \in M$ be arbitrarily fixed. By assumption (c) the MFCQ is fulfilled at $\tilde{x}$, i.e.

(MF1) $Dh_i(\tilde{x})$, $i \in I$, are linearly independent.

(MF2) There exists a vector $\xi := \xi(\tilde{x})$ with $Dh_i(\tilde{x})\xi = 0$, $i \in I$, and $Dg_j(\tilde{x})\xi < 0$ for all $j \in J_0(\tilde{x})$.

Let $t \in [0, 1]$ and $(\tilde{x}, \tilde{v}) \in M_1(t)$ be arbitrarily fixed. First, we have that the vectors

$$\begin{pmatrix} Dh_i(\tilde{x}) \\ 0 \end{pmatrix}, \qquad i \in I,$$

are linearly independent. Now we show that there exists an MF vector

$$\begin{pmatrix} \xi \\ \eta \end{pmatrix} := \begin{pmatrix} \xi \\ \eta \end{pmatrix} (\tilde{x}, \tilde{v}) \in \mathbb{R}^n \times \mathbb{R}.$$

We have to consider four cases:

(a) $K(\tilde{x}, \tilde{v}) := \{k \in \{1, \dots, L\} \mid \lambda_k^0 f_k(\tilde{x}) - \tilde{v} = \lambda_k^0 [\mu_k^0 + t(\mu_k^1 - \mu_k^0)]\} - \varnothing$,
  $J_0(\tilde{x}) = \varnothing$,

(b) $K(\tilde{x}, \tilde{v}) = \varnothing, \qquad J_0(\tilde{x}) \neq \varnothing$,

(c) $K(\tilde{x}, \tilde{v}) \neq \varnothing, \qquad J_0(\tilde{x}) = \varnothing$,

(d) $K(\tilde{x}, \tilde{v}) \neq \varnothing, \qquad J_0(\tilde{x}) \neq \varnothing$.

Case (a): (MF2) implies that the MFCQ is fulfilled at $(\tilde{x}, \tilde{v})$.

Case (b): We choose e.g. $\begin{pmatrix} \xi \\ \eta \end{pmatrix} = \begin{pmatrix} \xi \\ 0 \end{pmatrix}$ with $\xi = \xi(\tilde{x})$ given in (MF2)

Case (c): We choose e.g. $\begin{pmatrix} \xi \\ \eta \end{pmatrix} = \begin{pmatrix} 0 \\ 1 \end{pmatrix}$.

Case (d): We choose $\begin{pmatrix} \xi \\ \eta \end{pmatrix} = \begin{pmatrix} \xi \\ \eta^0 \end{pmatrix}$ with $\xi = \xi(\tilde{x})$ given in (MF2) and
  $\eta^0 > \max_{k \in \{1, \dots, L\}} \{\lambda_k^0 Df_k(\tilde{x})\xi\}$.

(2) As in the proof of Theorem B in [87] we use the proposed MF vectors for the construction of a vector field. We need that the MF vector depending on $(x, v) = (\tilde{x}, \tilde{v})$ does not vanish in a neighbourhood of a boundary point of $M_1(t)$. Since M is a compact set, this holds in the cases (b) and (d). The MF vector does not depend on (x, v) in case (c). Then we can follow the scheme of the proof in [87].

REMARK 6.4.3 We note that all assumptions of Theorem 6.4.1 are quite natural ones. Further, note that the theorem is valid for the feasible set $M_3(t)$ of the third parametrization, too. For the second parametrization $P_2(t)$, of course, the MFCQ is quite a natural assumption, because M is independent of t.

Now we can give the answer to Question 6.4.3 with the following remark.

REMARK 6.4.4   We distinguish between convex and non-convex problems and assume that

(W3) the MFCQ is fulfilled for all $x \in M$.

So, taking into account also (W1) and (W2), the comparison between the three parametrization from the point of view of the kind of singularities that could appear is given in Table 6.1.

Next, we give some comments on Table 6.1.

(i) In the convex case, $t = 1$ will be attained by pathfollowing methods only (without jumps) for the three parametrizations.

(ii) In the non-convex case the singularities that could appear for the three parametrizations lead to a solution algorithm (pathfollowing with jumps).

Summarizing, we can conclude that $P_1(t)$ and $P_3(t)$ can be solved by using pathfollowing methods (convex case) or pathfollowing methods with jumps (non-convex case) and with both parametrizations we have a chance to find a goal realizer (cf. Remark 6.4.2 (iii)), but the same kind of difficulties can appear as explained in Sections 6.2 and 6.3 (cf. $P^4(t)$, Figure 6.2), that is, since $M_i(t)$, $i = 1, 3$, are unbounded for all $t \in [0, 1]$, a $\bar{t} \in (0, 1)$ could appear, where $|v(t)|$ tends to $+\infty$ if t converges to $\bar{t}$. In this case, $t = 1$ will not be attained in a simple way. From this points of view, $P_1(t)$ and $P_3(t)$ have the same quality. However, $P_1(t)$ generates efficient points whereas $P_3(t)$ generates surely only weakly efficient points.

The parametrization $P_2(t)$ has a good behaviour for pathfollowing methods and $t = 1$ can always be attained by using algorithm JUMP I, but it is no guarantee to find a goal realizer even in the convex case.

Table 6.1

	CONVEX CASE	NON-CONVEX CASE
$P_1(t)$	Type 2, (b)* Type 5, (i)*	Type 2, (b), (c), (d)* Type 3, (e), (f)* Type 5, (i)*
$P_2(t)$	–	Type 2, (b), (c), (d)* Type 3, (e), (f)* Type 5, (i)*
$P_3(t)$	Type 2, (b)* Type 5, (i)*	Type 2, (b), (c), (d)* Type 3, (e), (f)* Type 5, (i)*

*In Figure 2.17.

References and Further Reading

[1] Alexander, J. C., Li, T.-Y. and Yorke, J. A., Piecewise smooth homotopies, in B. C. Eaves, F. J. Gould, H. O. Peithgen and M. J. Todd (eds), *Homotopy Methods and Global Convergence*, Plenum, New York, 1983.

[2] Allgower, E. L. and Georg, K., Simplical and continuation for approximating fixed points and solutions to systems of equations, *SIAM Rev.* **22** (1980), 28–85.

[3] Allgower, E. L. and Georg, K., Predictor–corrector methods for approximating fixed points and zero points of nonlinear mappings, in A. Bachem, M. Groetschel and B. Korte (eds), *Mathematical Programming, The State of the Art, Bonn 1982*, Springer-Verlag, Berlin, 1983, pp 15–57.

[4] Allgower, E. L. and Georg, K., *Introduction to Numerical Continuation Methods*, Springer-Verlag, Berlin, to appear.

[5] Allgower, E. L. and Schmidt, P. H., An algorithm for piecewise-linear approximation of an implicity defined manifold, *SIAM J. Numer. Anal.* **22** (1985), 322–46.

[6] Allgower, E. L. and Gnutzmann, S., An algorithm for piecewise linear approximation of implicitly defined 2-dimensional surfaces, *SIAM J. Numer. Anal.* **24** (1987), 452–69.

[7] Armacost, R. L. and Fiacco, A. V., Computational experience in sensitivity analysis for nonlinear programming, *Math. Program.* **6** (1974), 301–26.

[8] Arnold, V. I., Gusein-Zade, S. M. and Varchenko, A. N., *Singularities of Differentiable Maps I*, Birkhauser, Basel, 1985.

[9] Aubin, J. P. and Cellina, A., *Differential Inclusions*, Springer-Verlag, Berlin, 1984.

[10] Auslender, A., Regularity theorems in sensitivity with nonsmooth data, in [88], pp. 9–15.

[11] Avila, J. W., The feasibility of continuation methods for nonlinear equations, *SIAM J. Numer. Anal.* **11** (1974), 102–21.

[12] Bank, B., Guddat, J., Klatte, D., Kummer, B. and Tammer, K., *Non-linear Parametric Optimization*, Akademie-Verlag, Berlin, 1982.

[13] Bank, B., Mandel, R. and Tammer, K., Parametrische Optimierung und Aufteilungsverfahren, in [146], pp. 107–23.

[14] Bank, B. and Mandel, R., *Parametric Integer Optimization*, Math. Res., vol. 39, Akademie Verlag, Berlin, 1988.

[15] Beer, K., *Loesung grosser linearer Optimierungsaufgaben*, VEB Deutscher Verlag der Wissenschaften, Berlin, 1977.

[16] Ben-Tal, A., Melman, A. and Zowe, J., *Curved Search Methods for Unconstrained Optimization*, Report No. 88, Universität Bayreuth, Mathematisches Institut, 1988.

[17] Bernau, H., Interactive methods for vector optimization, in B. Brosowski and E.

Martinson (eds), *Optimization in Mathematical Physics*, Peter Lang Verlag, Frankfurt, 1986, pp. 21–37.

[18] Bertsekas, D., On penalty and multiplier methods for constrained minimization, *SIAM J. Control Optim.* **14** (1976), 216–35.

[19] Best, M. J., Braeuninger, J., Ritter, K. and Robinson, S. M., A globally and quadratically convergent algorithm for general nonlinear programming problems, *Computing* **26** (1981) 141–53.

[20] Bigelow, J. H. and Shapiro, N. Z., Implicit function theorems for mathematical programming, *Math. program.* **6** (1974) 142–56.

[21] Boot, J. C. G., On sensitivity analysis in convex quadratic programming problems, *Oper. Res.* **11** (1963), 771–86.

[22] Bröcker, Th. and Lander, L., *Differentiable Germs and Catastrophes*, Lond. Math. Soc. Lect. Notes Ser., Vol. 17, Cambridge University Press, 1975.

[23] Brosowski, B., A criterion of efficiency and some application in mathematical physics, in B. Brosowski and E. Martinson (eds), *Optimization in Mathematical Physics*, Peter Lang Verlag, Frankfurt, 1986, pp. 37–61.

[24] Brosowski, B., *Parametric Semi-Infinite Optimization*, Peter Lang Verlag, Frankfurt, 1982.

[25] Brosowski, B. and Conci, A., On vector optimization and parametric programming, in *Proc. Conf. Segundas Jornadas Latino Americas de Matematica Aplicada, Rio de Janeiro 1983*, Vol. II, pp. 483–95.

[26] Brosowski, B., On the continuity of the optimum set in parametric semiinfinite programming, in A. V. Fiacco (ed.), *Mathematical Programming with Data Perturbations II*, Lect. Notes Pure Appl. Math. 85, Marcel Dekker, New York, 1983, pp. 23–49.

[27] Brosowski, B. and Deutsch, F. (eds), *Parametric Optimization and Application*, Conference Held at the Mathemat. Forschungsinstitut Oberwolfach, 16–22 Oct. 1983, Int. Ser. Numer. Math., Birkhauser, Basel, 1985.

[28] Burghard, S. and Richter, C., Ein Praediktor–Korrektor–Verfahren der nichtlinearen Optimierung, *Wissensch. Z. Techn. Univ. Dresden* **31** (1982), 193–8.

[29] Chandrasekaran, R., A special case of the complementarity pivot problem, *Oper. Res.* **7** (1970), 263–8.

[30] Charnes, A. and Cooper, W., *Management Models and Industrial Applications of Linear Programming*, Wiley, New York, 1961.

[31] Charnes, A., Garcia, C. B. and Lemke, C. E., Constructive proofs of theorems relating to $F(x) = y$, with applications, *Math. Program.* **12** (1977), 328–43.

[32] Charnes, A. and Zlobec, S., Stability of efficiency evaluations in data envelopment analysis, *Z. Oper. Res., Ser. A: Theor.*, to appear.

[33] Chow, S. N., Mallet-Paret, J. and Yorke, J. A., Finding zeros of maps: homotopy methods that are constructive with probability one, *Math. Comput.* **32** (1978), 887–99.

[34] Den Heijer, C. and Rheinboldt, W. C., On steplength algorithms for a class of continuation methods, *SIAM J. Numer. Anal.* **18** (1981), 925–48.

[35] Deutsch, F. and Kenderov, P., Continuous selections and approximate selections for set-valued mappings and applications to metric projections, *SIAM J. Math. Anal.* **14** (1983), 185–94.

[36] Diener, I., On global convergence of path-following methods to determine all solutions to a system of nonlinear equations, *Math. program.* **39** (1987), 181–9.

[37] Diener, I., Trajectory nets connecting all critical points of a smooth function, *Math. Program.* **36** (1986), 340–52.

[38] Dinkelbach, W., *Sensitivitaetsanalysen und parametrische Programmieruno*, Springer-Verlag, Berlin, 1969.

[39] Dommisch, G., Existence of Lipschitz-continuous and differentiable selections for multifunctions, in [88], pp. 60–73.

[40] Dontchev, A. L., *Perturbations, Approximations and Sensitivity Analysis of Optimal Control Systems*, Lect. Notes Control Inform. Sci., Springer-Verlag, Berlin, 1983.

[41] Dontchev, A. L. and Jongen, H. Th., On the regularity of the Kuhn–Tucker curve, *SIAM J. Control Optim.* **24** (1986), 169–76.

[42] Dupacova, J., On some connections between parametric and stochastic programming, in [88], pp. 74–81.

[43] Eaves, B. C., The linear complementarity problem, *Management Sci.* **17** (1971), 612–34.

[44] Eaves, B. C., A short course in solving equations with PL homotopies, in R. W. Cottle and C. E. Lemke (eds), *Nonlinear programming*, Proc. Ninth SIAM–AMS Symp. in Applied Mathematics, SIAM, Philadelphia, 1976, pp. 73–143.

[45] Eaves, B. C. and Scarf, H., The solution of systems of piecewise linear equations, *Math. Oper. Res.* **1** (1976), 1–27.

[46] Eremin, I. I. and Guddat, J. (eds), *Parametric Optimization and Ill-posed Problems in Mathematical Optimization*, Seminarberich No. 81, Sektion Mathematik der Humboldt-Universität zu Berlin, 1986.

[47] Ester, J. and Tröltzsch, F., On generalized notions of efficiency in multicriteria decision making, *Syst. Anal. Model. Simul.* **2** (1986), 147–55.

[48] Fiacco, A. F., Sensitivity analysis for nonlinear programming using penalty functions, *Math. Program.* **10** (1976), 287–311.

[49] Fiacco, A. F., *Introduction to Sensitivity and Stability Analysis in Nonlinear Programming*, Academic Press, New York, 1983.

[50] Fiacco, A. V. and Kyparisis, J., Sensitivity analysis in nonlinear programming under second order assumptions, in A. Bagchi and H. Th. Jongen (eds), *Systems and Optimization*, Lect. Notes Control Inform. Sci. 66, Springer-Verlag, Berlin, 1985, pp. 74–97.

[51] Fiacco, A. V. and Kyparisis, J., Computable bounds on parametric solutions of convex problems, *Math. Program.* **40** (1980), 213–21.

[52] Fiacco, A. V. and McCormick, G. P., *Nonlinear Programming: Sequential Unconstrained Minimization Techniques*, Wiley, New York, 1968.

[53] Fischer, T., Continuous selections for semi-infinite optimisation, in [88], pp. 95–112.

[54] Fletcher, R., *Practical Methods of Optimization*, Vol. 2, *Constrained Optimization*, Wiley, New York, 1981.

[55] Fujiwara, O., A note on differentiability of global optimal values, *Math. Oper. Res.* **10** (1985), 612–18.

[56] Fukushima, M., Solving inequality constrained optimization problems by differential homotopy continuation methods, *J. Math. Anal. Applic.* **133** (1988), 109–21.

[57] Gal, T., *Betriebliche Entscheidungsprobleme, Sensitivitaetsanalyse und parametrische Programmierung*, Walter der Gruyter, Berlin, 1973.

[58] Gal, T., *Post-optimal Analysis, Parametric Programming Analysis and Related Topics*, McGraw-Hill, New York, 1979.

[59] Gal, T., On efficient sets in vector maximum problems—a brief Survey, in P. Hansen (ed.), *Essays and Surveys on Multiple Criteria Decision Making*, Lect. Notes Econ. Math. Syst. 209, Springer-Verlag, Berlin, 1982.

[60] Gantmacher, F. R., *Matrizenrechnung*, Teil I, VEB Deutscher Nerlag der Wissenschaften, Berlin, 1970.

[61] Garcia-Palomares, U. M. and Mangasarian, O. L., Superlinearly convergent quasi-Newton algorithms for nonlinearly constrained optimization problems, *Math. Program.* **11** (1976), 1–13.

[62] Garcia, C. B. and Zangwill, W. I., *Pathways to Solutions, Fixed Points and Equilibria*, Prentice-Hall, Englewood Cliffs, NJ, 1981.

[63] Barcia, C. B. and Gould, F. J., An application of homotopy to solving linear programs, *Math. Program.* **27** (1983), 263–82.

[64] Gauvin, J., A necessary and sufficient regularity condition to have bounded multipliers in nonconvex programming, *Math. Program.* **12** (1977) 136–8.

[65] Geoffrion, A. M., Strictly concave parametric programming. Part I: Basic theory, *Management Sci., Ser. A* **13** (1966), 244–53. Part II, Additional theory and computational considerations, *Management Sci., Ser. A* **13** (1967), 359–70.

[66] Geoffrion, A. M., Generalized Bender's decomposition, *JOTA* **10** (1972), 273–59.

[67] Georg, K, On tracing an implicitly defined curve by Quasi-Newton steps and calculating bifurcation by local perturbation, *SIAM J. Stat. Comp.* **2** (1981), 35–49.

[68] Gfrerer, H., Hölder continuity of solutions of perturbed optimization problems under Mangasarian-Fromovitz constraint qualification, in [88], pp. 113–24.

[69] Gfrerer, H., Globalisierung der Multiplikatormethode in der nichtlinearen Optimierung mit Hilfe von Einbettung, Dipl. Arbeit, Math. Inst., Universität Linz, 1982.

[70] Gfrerer, H., Guddat, J. and Wacker, J., A globally convergent algorithm based on imbedding and parametric optimizition, *Computing* **30** (1983), 225–52.

[71] Gfrerer, H., Guddat, J., Wacker, J. and Zulehner, W., Path-following methods for Kuhn–Tucker curves by an active index set strategy, in A. Bagchi, and H. Th. Jongen (eds), *System and Optimization*, Proc. Twente Workshop, Lect. Notes Control Inform. Sci. 66, Springer-Verlag, Berlin, 1985, pp. 111–32.

[72] Giannessi, F., Theorems of the alternative and optimality conditions, *J. Optim. Theory. Applic.* **42** (1984), 331–65.

[73] Gollmer, R., On linear multiparametric optimization with parameter-dependent constraint matrix, *Optimization* **16** (1985), 15–28.

[74] Golikov, A. I. and Kotkin, G. G., *Application of Sensitivity Function in Multicriteria Optimization* (in Russian), Moscow, 1986.

[75] Golubitsky, M. and Schaeffer, D. G., *Singularities and Groups in Bifurcation Theory I*, Springer-Verlag, Berlin, 1985.

[76] Grauer, M., Lewandowski, A. and Wierzbicki, A. P., DIDAS: theory, implementation and experience, in M. Grauer and A. P. Wierzbicki (eds), *Interactive Decision Analysis*, Lect. Notes Econ. Math. Syst. 229, Springer-Verlag, Berlin, 1984.

[77] Grossmann, Ch. and Kleinmichel, H., *Verfahren der nichtlinearen Optimierung*, Teubner-Verlagsgeselischaft, Leipzig, 1976.

[78] Grossmann, Ch. and Kaplan, A. A., *Strafmethoden und modifizierte Lagrangemethoden in der nichtlinearen Optimierung*, Teubner-Verlagsgesollschaft, Leipzig, 1979.

[79] Grygerova, L., Loesungsbereich von Optimierungsproblemen mit Parametern in den Koeffizienten der Matrix der linearen Restriktionsbedingungen, *Apl. Mat.* **17** (1972), 388–400.

[80] Guddat, J., Parametric optimization: pivoting and predictor-corrector continuation, a survey, in [88], 309–63.

[81] Guddat, J. and Guerra Vasquez, F., Multiobjective optimization using pivoting and continuation methods, *Arch. Autom. Telemech.* 301–18.

[82] Guddat, J., On globally convergent algorithms: singularities, pathfollowing and jumps. Preprint, Universita di Pisa, Dipartimento di Matematica, to appear, 1990.

[83] Guddat, J., Guerra Vasquez, F., Tammer, K. and Wendler, K., Multi-objective and stochastic optimization based on parametric optimization, Math. Research Vol. 26, Akademic-Verlag, Berlin, 1985.

[84] Guddat, J., Guerra Vasquez, F., Tammer, K. and Wendler, K., On dialogue algorithms of multiobjective optimization problems based on parametric optimization, in [46], pp. 16–69.

[85] Guddat, J. and Jongen, H. Th., On global optimization based on parametric optimization, in J. Guddat *et al.* (eds), Advances in Math. Optimization. Math. Res., Vol. 45, Akademic-Verlag, Berlin, 1988, pp. 63–79.

[86] Guddat, J. and Jongen, H. Th., Structural stability in nonlinear optimization, *Optimization* **18** (1987), 617–31.

[87] Guddat, J., Jongen, H., Th. and Rüeckmann, J., On stability and stationary points in nonlinear optimization, *J. Aust. Math. Soc., Ser. B* **28** (1986), 36–56.

[88] Guddat, J., Jongen, H., Th., Kummer, B. and Nožička, F. (eds), *Parametric Optimization and Related Topics*, Math. Res. Vol. 35, Akademic-Verlag, Berlin, 1987.

[89] Guddat, J., Nowack, D., Rückmann, J. and Ruske, A., On singularities and pathfollowing methods with jumps in parametric optimization and application in multiobjective optimization, IIASA-Working Paper (to appear).

[90] Guddat, J., Jongen, H., Th. and Nowack, D., Parametric optimization pathfollowing with jumps, in A. Gomez, F. Guerra, M. A. Jimenez and G. Lopez (eds), *Approximation and Optimization*, Lect. Notes Math. 1354, Springer-Verlag, Berlin, 1988, pp. 43–53.

[91] Guddat, J., Jongen, H. Th. and Nowack, D., Parametric optimization: pathfollowing and jumps in the set of local minimizers and in the critical set, in [246] to appear.

[92] Guddat, J. and Tammer, K., Eine Modifikation der Methode von Theil und van de Panne zur Loesung einparametrischer quadratischer Optimierungsproblemme, *Math. Operationsf. Stat.* **1** (1970), 199–206.

[93] Guddat, J., Wacker, Hj. and Zulehner, W., On imbedding and parametric optimization—a concept of a globally convergent algorithm for nonlinear optimization problems, *Math. Program. Stud.* **21** (1984), 79–96.

[94] Guerra, F. Algunas posibilidades de utilización de la optimización paramétrica en la optimización vectorial cuadrática *Revista Investigación Operacional*, **2** (1981), 2–3.

[95] Gustafson, S.-A. and Kortanek, K. O., Semi-infinite programming and applications, in A. Bachem, M. Groetschel and B. Korte (eds), *Mathematical Programming, The State of Art*, Springer-Verlag, Berlin, 1983, pp. 132–57.

[96] Hackl, J., Solution of optimization problems via continuation methods, in [222], pp. 95–127.

[97] Hackl, J., Wacker, Hj. and Zulehner, W., and efficient stepsize control for continuation methods, *Bit* **20** (1980), 475–85.

[98] Han, S.-P. and Fujiwara, O., An inertia theorem for symmetric matrices and its applications to nonlinear programming. *Lin. Alg. Appl.* **72** (1985), 47–58.

[99] Houthakker, S., The capacity method of quadratic programming, *Econometrica* **28** (1960), 62–87.

[100] Hestenes, M. R., Augmentability in optimization theory, *JOTA* **32** (1980), 427–40.

[101] Hettich, R. and Jongen, H. Th., Semiinfinite programming: conditions of optimality and applications, in *Optimization Techniques*, Part 2, Lect. Notes Control Inf. Sci. 7, Springer-Verlag, Berlin, 1978, pp. 1–11.

[102] Hettich, R. and Jongen, H. Th., On the local continuity of the Chebyshev operator, *JOTA* **33** (1981), 296–307.

[103] Hettich, R and Still, G., Local aspects of a method for solving membrane-eigenvalue problems by parametric semi-infinite programming, in [88].

[104] Hettich, R. and Zencke, P., *Numerische Methoden zur Approximation und semi-infiniten Optimierung*, Teubner Studienbücher, Stuttgart, 1982.

[105] Hirsch, M. W., *Differential Topology*, Grad. Texts Math., Vol. 33, Springer-Verlag, Berlin, 1976.

[106] Hock, W. and Schittkowski, K., A comparative performance evaluation of 27 nonlinear programming codes, *Computing* **30** (1983), 335–58.

[107] Hogan, W. W., Point-to-set maps in Mathematical programming, *SIAM Rev.* **15** (1973), 519–603.

[108] Huneault, M., Calderon, R. and Galiana, F. D., Fast secure economic dispatch using continuation methods, Preprint, Department of Electrical Engineering, McGill University, Montreal, Quebec, 1984.

[109] Jadikin, A. B., Nonlinear parametric optimization of large-scale systems (in Russian), *Sbornik VNI* **13** (1984), 55–67.

[110] Jarre, F., *On the Convergence of the Method of Analytic Centres when applied to Convex Quadratic Programs*, Report No. 35, Universität Wuerzburg, 1987.

[111] Jittorntrum, K., Solution point differentiability without strict complementarity in nonlinear programming, *Math. Program.* **21** (1984), 127–38.

[112] Jongen, H. Th., Parametric optimization: critical points and local minima, in *Proc. Seminar on Computational Solution of Nonlinear Systems of Equations* (Colorado State University, 18–29 July 1988), Preprint No. 2, 1988, Lehrstuhl für Mathematick, RWTH Aachen, West Germany, 1988; to appear in *Lect. Appl. Math.*

[113] Jongen, H. Th., Three lectures on nonlinear optimization, in P. Kenderov (ed.), *Mathematical Methods in Operations Research*, Bulgarian Academy of Sciences, 1985, pp. 60–8.

[114] Jongen, H. Th., Jonker, P. and Twilt, F., On deformation in optimization, *Meth. Oper. Res.* **37** (1980), 171–84.

[115] Jongen, H. Th., Jonker, P. and Twilt, F., On one-parameter families of sets defined by (in) equality constraints, *Nieuw Arch. Wiskunde (3)*, **30** (1982), 307–22.

[116] Jongen, H. Th., Jonker, P. and Twilt, F, On one-parametric families of optimization problems: equality constraints, *J. Optim. Theor. Applic.* **48** (1986), 141–61.

[117] Jongen, H. Th., Jonker, P. and Twilt, F., *On Index-sequence Realization in Parametric Optimization*, Seminarbericht No. 50 der Sektion Mathematik der Humboldt-Universität zu Berlin, 1983, pp. 159–66.

[118] Jongen, H. Th., Jonker, P. and Twilt, F., Critical sets in parametric optimization, *Math. Program.* **34** (1986), 333–53.

[119] Jongen, H. Th., Jonker, P. and Twilt, F., Parametric optimization: the Kuhn-Tucker set, in [88], pp. 196–208.

[120] Jongen, H. Th., Jonker, P. and Twilt, F., *Nonlinear Optimization in $\mathbb{R}^n$. I. Morse Theory, Chebishev Approximation*, Peter Lang Verlag, Frankfurt, 1983.

[121] Jongen, H. Th., Jonker, P. and Twilt, F., *Nonlinear optimization in $\mathbb{R}^n$. II. Transversality, Flows, Parametric Aspects*, Peter Land Verlag, Frankfurt, 1983.

[122] Jongen, H. Th., Klatte, D. and Tammer, K., *Implicit Functions and Sensitivity of*

Stationary Points, Preprint No. 1, Lehrstuhl für Mathematik, RWTH Aachen, 1988; to appear in *Math. Program.*

[123] Jongen, H. Th., Moebert, T. and Tammer, K., On iterated minimization in nonconvex optimization, *Math. Oper. Res.* **11** (1986), 679–91.

[124] Jongen, H. Th. and Zwier, G., On the local structure of the feasible set in semi-infinite optimization, *Int. Ser. Numer. Math.* **72** (1985), 185–202.

[125] Jongen, H. Th. and Zwier, G., Structural analysis in semi-infinite optimisation, in Lemarechal C. (ed.), *Proc. Third Franco-German Conf. in Optimization*, INRIA 1985, pp. 56–67.

[126] Jongen, H. Th. and Zwier, G., On regular semi-infinite optimization, in E. J. Anderson and A. B. Philpott (eds), *Infinite Programming*, Lect. Notes Econ. Math. Syst. 259, Springer-Verlag, Berlin, 1985, pp. 53–64.

[127] Kall, P., On approximation in stochastic programming, in [88], pp. 387–407.

[128] Kelley, J. K., *General Topology*, Van Nostrand-Reinhold, New York, 1969.

[129] Klatte, D. and Kummer, B., Stability properties of infima and optimal solutions of parametric optimization problems, in *Abstracts of the IIASA Workshop on Nondifferentiable Optimization Motivations and Applications*, 17–22 Sept. 1984, Sorpron, Hungary, IIASA, Laxenburg, Austria, 1984.

[130] Klatte, D., Lipschitz continuity of infima and optimal solutions in parametric optimization: the polyhedral case, in [88], pp. 229–49.

[131] Kleinmann, P. and Schultz, R., A simple procedure for optimal load dispatch using parametric programming, *Z. Oper. Res.* (to appear).

[132] Kojima, M., A complementary pivoting approach to parametric nonlinear programming, *Meth. Oper. Res.* **4** (1979), 464–72.

[133] Kojima, M., On the homotopic approach to systems of equations with separable mappings, *Math. Program. Stud.* **7** (1978), 170–84.

[134] Kohima, M., Strongly stable stationary solutions in nonlinear programs, in *Analysis and Computing of Fixed Points*, Academic Press, and New York, 1980, pp. 93–138.

[135] Kojima, M. and Hirabayashi, P., Continuation deformation of nonlinear programs, *Math. Program. Stud.* **21** (1984), 150–98.

[136] Kojima, M., Nishino, H. and Sekine, T., An extension of Lemke's method to the piecewise linear complementarity problem, *SIAM J. Appl. Math.* **31** (1976), 600–13.

[137] Kuhn, H. W. and Tucker, A. W., Nonlinear programming, in J. Neymann (ed.), *Proc. Second Berkeley Symp. on Math. Statistics and Probability*, Berkeley, Calif., Univ. of California, 1951, pp. 481–92.

[138] Kummer, B., Linearly and nonlinearly perturbed optimization problems, in [88], pp. 249–68.

[139] Kummer, B., *The Inverse of a Lipschitz Function in $\mathbb{R}^n$: Complete Characterization by Directional Derivatives*, Preprint, Humboldt University, Berlin, Department of Mathematics, Berlin, 1988.

[140] Lehmann, R., On the numerical feasibility of continuation methods for nonlinear programming problems, *Math. Oper. Forsch. Stat. Ser. Opt.* **15** (1984), 517–30.

[141] Lehmann, R., An algorithm for solving one-parametric optimization problems based on an active-index set strategy, in [88], pp. 268–301.

[142] Lemke, C. E., Bimatrix equilibrium points and mathematical programming, *Management Sci.* **11** (1965), 681–9.

[143] Levitin, E. S., On corrections of solutions of nonlinear optimization problems with incorrect information (in Russian), *Akad. Nauk SSSR, Summer school*, Irkutsk, 1974.

[144] Ljusternik, L. A. and Sobolew, W. I., *Elemente der Funktionalanalysis*, Akademie-Verlag, Berlin, 1968.

[145] Loridan, P. and Morgan, J., New results on approximate solutions in two-level optimization, *Optimization* to appear.

[146] Lommatzsch, K. (ed.), *Anwendungen der linearen parametrischen Optimierung*, Akademie-Verlag, Berlin, 1979.

[147] Luenberger, D. G., *Introduction to Linear and Nonlinear Programming*, Addison-Wesley, London, 1973.

[148] Lüthi, H. J., *Komplementaritaets—und Fixpunktalgorithmen in der Mathematischen Programmierung*, Lect. Notes Econ. Math. Syst., 129, Springer-Verlag, Berlin, 1976.

[149] Malanowski, K., *Stability of Solutions to Convex Problems of Optimization*, Lect. Notes Control Inform. Sci. 93, Springer-Verlag, Berlin, 1987.

[150] Malanowski, K., Differentiability with respect to parameters, *Math. Program.* **33** (1985), 352–61.

[151] Malanowski, K., Stability and sensibility to optimal control problems for systems with control appearing linearly, *Appl. Math. Optim.* **16** (1987), 73–91.

[152] Malanowski, K., Higher order sensitivity of solutions to convex programming problems without strict complementarity, in M. Iri and Y. Yajima (eds), *Proc. 13th IFIP Conf. on System Modelling and Optimization*, Tokyo, 31 Aug.–4 Sept. 1987, Lect. Notes Control Inform. Sci., Springer-Verlag, Berlin (to appear).

[153] Mangasarian, O. L., Equivalence of the complementarity problem to a system of nonlinear equations, *SIAM J. Appl. Math.* **31** (1976) 1.

[154] Mangasarian, O. L. and Fromovitz, S., The Fritz John necessary optimality conditions in the presence of equality and Inequality constraints, *J. Math. Anal. Applic.* **17** (1967) 37–47.

[155] Marcus, M. and Minc, H., *A Survey of Matrix Theory and Matrix Inequalities*, Allyn and Bacon, Boston, 1964.

[156] Matsumoto, T., Shindoh, S. and Hirabayashi, R., *A New Characterization of Mangasarian-Fromovitz Condition*, Preprint, Research Reports on Information Sciences, Series B, Operations Research, No. B-214, Tokai Regional Fisheries Research Laboratory, Tokyo, 1988.

[157] Matsumoto, T., Shindoh, S. and Hirabayashi, R., *1-determinacy of Feasible Sets*, Preprint. Research Reports on Information Sciences, Series B, Operations Research, No. B-215, Tokai Regional Fisheries Research Laboratory, Tokyo, 1988.

[158] McCormick, G. P., *Nonlinear Programming: Theory, Algorithms, and Applications*, Wiley, New York, 1983.

[159] Megiddo, N., On the parametric nonlinear complementarity problem, *Math. Program. Stud.* **7** (1978), 142–50.

[160] Megiddo, N. and Kojima, M., On the existence and uniqueness of solutions in nonlinear complementarity theory, *Math. Program.* **12** (1977), 110–30.

[161] Meravy, P., A note about some relations between the method of parametric transformation functions on the smooth-homotopy methods for solving constrained optimization problems, *Proc. 8th Conf. Math. Meth. in Economy*, Sellin, GDR, 1984.

[162] Meravy, P., Smooth homotopies for mathematical programming, in [88], pp. 302–15.

[163] Michael, E., Selected selection theorems, *Am. Math. Monthly* **63** (1956), 230–7.

[164] Milnor, J., Lectures on the *h*-cobordism theorem, *Math. Notes* 1, Princeton University Press, 1965.

[165] Milnor, J., Morse theory, *Ann. Math. Stud.*, no. 51, Princeton University Press, 1963.

[166] Mörwald, J., Loesung nichtlinearer, einparametrischer Optimierungsprobleme mit einer aktiven Indexmengenstrategie, Diplomarbeit, Institut für Mathematik, Kepler-Universität, Linz, 1985.

[167] Muu, L. and Dettli, W., *An Algorithm for Indefinite Quadratic Programming with Convex Constraints*, No. 89, Fakultät für Mathematik und Informatik, Universität Mannheim, 1989.

[168] Nožička, F., Guddat, J. Hollatz, H. and Bank, B., *Theorie der linearen parametrischen Optimierung*, Akademie-Verlag, Berlin, 1974.

[169] Ortega, J. M. and Rheinboldt, W. C., *Iterative Solutions of Nonlinear Equations in Several Variables*, Academic Press, New York, 1970.

[170] van de Panne, C., *Methods for Linear and Quadratic Programming*, North-Holland, Amsterdam, 1975.

[171] Pappalardo, M., *A Generalization of Penalty Methods via Image Problem*, Preprint, Università di Pisa, Dipartimento di Matematica, 1986

[172] Pareto, V., *Course d'Economic Politique*, Rouge, Lausanne, 1986.

[173] Pateva, D., Strukturuntersuchungen Für lineare einparametrische Optimierungsaufgaben, Diss., Humboldt Univ. Berlin, Sekt. Mathematik, 1989.

[174] di Pillo, G. and Grippo, L., A continuously differentiable exact penalty function for nonlinear programming problems with inequality constraints, *SIAM J. Control Optim.* **23** (1985), No. 1.

[175] Poenisch, B. and Schwetlick, H., Computing turning points of curves implicitly defined by nonlinear equations depending on a parameter, *Computing* **26** (1981), 107–21.

[176] Poore, A. B. and Tiahrt, C. A., Bifurcation problems in nonlinear parametric programming, *Math. Program.* **39** (1987), 189–206.

[177] Reinoza, A., Solving generalized equations via homotopies, *Math. Program.* **31** (1985), 307–20.

[178] Richter, C., *Ueber die numerische Behandlung von Einbettungsverfahren in der nichtlinearen Optimierung*, TU Dresden, Preprint, 1980.

[179] Richter, C., Ein implementierbares Einbettungsverfahreh der nichtlinearen Optimierung, *Math. Operationsf. Stat. Ser. Optim.* **15** (1984), 545–53.

[180] Richter, C., Ueber die numerische Behandlung nichtlinearer Optimierungsprobleme mit Hilfe von verallgemeinerten Variationsungleichungen und von Nichtoptimalitaetsmassen, Diss. B, Technische Universitaet Dresden, 1980.

[181] Richter, C., Zur globalen Konvergenz des gedaempften Wilson-Verfahrens. *Math. Operationsf. Stat. Ser. Optim.* **10** (1979), 213–18.

[182] Richter, C., Numerical methods for solving parametric optimization problems. Submitted for IIASA-Conference Report, Irkutsk, 1989.

[183] Ritter, K., A method for solving, maximum-problems with a nonconcave quadratic objective function, *Z. Wahrsch. Verw. Gebiete* **4** (1965), 340–51.

[184] Ritter, K., A method for solving maximum-problems depending on parameters, *Naval Res. Logist. Quart.* **14** (1967), 147–62.

[185] Ritter, K., A parametric method for solving certain nonconcave maximum problems, *J. Computer Syst. Sci.* **1** (1967), 44–54.

[186] Ritter, K., Ein Verfahren zur Loesung parameterabhaengiger Maximumprobleme, *Unternehmungsforschung* **6** (1962) 149–96.

[187] Robinson, S. M., A quadratically-convergent algorithm for general nonlinear programming problems, *Math. Program.* **3** (1972), 145–56.

[188] Robinson, S. M., *An Implicit-function Theorem for Generalized Variational Inequalities*, MRC Technical Summary Report 1672, University of Wisconsin, Madison, 1976.

[189] Robinson, S. M., Perturbed Kuhn–Tucker points and rates of convergence for a class of nonlinear programming algorithms, *Math. Program.* 7 (1974), 1–16.

[190] Robinson, S. M., Strongly regular generalized equations, *Math. Oper. Res.* 5 (1980), 43–62.

[191] Robinson, S. M. (ed.), *Analysis and Computation of Fixed Points*, Academic Press, New York, 1980.

[192] Robinson, S. M., Generalized equations and their solutions, Part I: Basic theory, *Math. Program. Stud.* 10 (1979), 12B–41.

[193] Robinson, S. M., Generalized equations and their solutions, Part II: Applications to nonlinear programming, *Math. Program. Stud.* 19 (1982), 200–21.

[194] Robinson, S. M., Stability theory for systems of inequalities. Part II: Differentiable nonlinear systems, *SIAM J. Numer. Anal.* 13 (1976), 497–513.

[195] Robinson, S. M., Generalized equations, in A. Bachem, M. Groetschel and B. Korte (eds), *Mathematical Programming, The State of the Art*, Springer-Verlag, Berlin, 1983, pp. 346–67.

[196] Robinson, S., *An Implicit-Function Theorem For B-Differentiable Functions*, Working Paper, July 1988, WP-88-67, IIASA, Laxenburg, Austria.

[197] Robinson, S. M., Local structure of feasible sets in non-linear programming. Part I: Regularity, in V. Pereyra and A. Reinoza (eds), *Numerical Methods*, Lect. Notes Math. 1005, Springer-Verlag, Berlin, 1983, pp. 240–51.

[198] Robinson, S., Local structure of feasible sets in nonlinear programming, Part III: Stability and sensitivity, *Math. Program. Stud.* 30 (1987), 45–66.

[199] Rückmann, J., Einparametrische nichtkonvexe Optimierung: Strukturuntersuchungen und eine Verallgemeinerung des Einbettungsprinzips, Dissertation, TH Leipzig, 1988.

[200] Rupp, Th., Kuhn–Tucker curves for one-parametric semi-infinite programming, *Optimization* 20 (1989), 61–77.

[201] Rupp, Th., Kontinuitaetsmethoden zur Loesung einparametrischer semi-infiniter Optimierungsprobleme, Dissertation, Universität Trier, 1988.

[202] Ruske, A., Numerical treatment of bifurcation problems in case of one-parametric nonlinear optimization problems, in [46], pp. 109–30.

[203] Ruske, A., Numerische Behandlung von Bifurkationsproblemen bei einparametrischen Optimierungsaufgaben, Dissertation, Sektion Mathematik der Humboldt-Universität zu Berlin, 1986.

[204] Ruszczynski, A., A regularized decomposition method for minimizing a series of polyhedral functions, *Math. Program.* 35 (1986), 309–33.

[205] Saigal, R., A note on a special linear complementarity problem, *Oper. Res.* 7 (1970), 179–83.

[206] Schecter, S., Structure of the first-order solution set for a class of nonlinear programs with parameters, *Math. Program.* 34 (1986), 362–9.

[207] Schittkowski, K., *Nonlinear Programming Codes–Information, Tests, Performance*, Lect. Notes Econ. Math. Syst. 183, Springer-Verlag, Berlin, 1980.

[208] Schittkowski, K., The nonlinear programming method of Wilson, Han and Powell with an augmented Lagrangian. Part 1: Convergence analysis, *Numer. Math.* 38 (1981), 83–114; Part 2: An efficient implementation with linear least squares subproblems, *Numer. Math.* 38 (1981), 115–27.

[209] Schwetlick, H., *Numerische Loesung nichtlinearer Gleichungen*, VEB Deutscher Verlag der Wissenschaften, Berlin, 1979.

[210] Semple, J. and Zlobec, S., On the continuity of a Lagrangian multiplier function in input optimization, *Math. Program.* **34** (1986), 362–9.

[211] Siersma, S., Singularities of functions on boundaries, corners, etc. *Q. J. Math. Oxford Ser. (2)* **32** (1981), 119–27.

[212] Shindoh, S., Hirabayashi, R. and Matsumoto, T., Structure of solution set to nonlinear programs with two parameters: I. Change of stationary indices, *Math. Program.* (1989), 8–224.

[213] Smale, S., Global analysis and economics V. *J. Math. Econ.* **1** (1974), 213–21.

[214] Sternberg, S., *Lectures on Differential Geometry*, Prentice-Hall, Englewood Cliffs, NJ, 1964.

[215] Tammer, K., Die Abhaengigkeit eines quadratischen Optimierungsprobleme von einem Parameter in der Zielfunktion, *Math. Operationsf. Stat.* **5** (1974), 573–90.

[216] Tammer, K., Möglichkeiten zur Anwendung der Erkenntnisse der parametrischen Optimierung, *Math. Operationsf. Statist.* **7** (1976), 209–222.

[217] Tammer, K., Relations between stochastic and parametric programming in decision problems with a random objective function, *Optimization* **9** (1978), 523–535

[218] Tammer, K., The application of parametric optimization and imbedding for the foundation and realization of a generalized primal decomposition approach, in [88], pp. 376–86.

[219] Tapia, R. A., Diagonalized multiplier methods and quasi-Newton methods for constrained optimization, *JOTA* **22** (1977), 135–94.

[220] Todd, M. J., New fixed-point algorithms for economic equilibria and constrained optimization, *Math. Program.* **18** (1980), 111–26.

[221] Välialo, H., A unified approach to une parametric general quadratic programming, *Math. Program.* **33** (1985), 318–38.

[222] Wacker, Hj. (ed.), *Continuation Methods*, Academic Press, New York, 1978.

[223] Wacker, Hj., A summary of the developments of imbedding methods, in [222], pp. 1–35.

[224] Watson, T., Solving the nonlinear complementary problem by a homotopy method, *SIAM J. Control Applic.* **17** (1979), 36–46.

[225] Wetterling, W., Definitheitsbedingungen fuer relative Extrema bei Optimierungs- und Approximationsaufgaben, *Numer. Math.* **15** (1970), 122–36.

[226] Wierzbicki, A. P., Basic properties of scalarizing functional for multiobjective optimization, *Math. Oper. Stat. Optim.* **8** (1977), 55–60.

[227] Wierzbicki, A. P., On the completeness and constructiveness of parametric characterizations to optimization problems, *OR Spectrum* **81** (1986), 73–87.

[228] Wierzbicki, A. P., Note on the equivalence of Kuhn–Tucker complementarity conditions to an equation, *JOTA* **37** (1982), 401–5.

[229] Wilson, R. B., A simplicial method for concave programming, Ph.D. Diss., Harvard Univ., 1963.

[230] Wolfe, Ph., The simplex-method for quadratic programming, *Econometrica* **27** (1959), 382–98.

[231] Yomdin, Y., The geometry of critical and near-critical values of differentiable mappings, *Math. Ann.* **264** (1983), 495–515.

[232] Zangwill, W. I., *Fixed Points, Equilibria and Homotopies*, Report 8102. Graduate School of Business, University of Chicago, 1981.

[233] Zangwill, W. I. and Garcia, C. B., Equilibrium programming: the path following approach and dynamics, *Math. Program.* **21** (1981), 262–89.

[234] Zhadan, V. G., A method for the parametric representation of objective functions

in conditional multicriterial optimization, *USSR Comput. Math. Phys.* **26** (1986), 108–15.

[235] Zlobec, S. and Ben-Israel, A., Perturbed convex-programs: continuity of optimal solutions and optimal values, *Oper. Res. Verf.* **31** (1979), 737–49.

[236] Zlobec, S., Characterizing an optimal input in perturbed convex programming, *Math. Program.* **25** (1983), 109–21.

[237] Zlobec, S., Input optimization I: Optimal realizations of mathematical models, *Math. Program.* **31** (1985), 245–68.

[238] Zlobec, S., Input optimization II: Optimal realizations of multiobjective models, *Optimization* **17** (1986), 429–45.

[239] Zlobec, S., Survey on input optimization, *Optimization* **18** (1987), 309–48.

[240] Zlobec, S., Stable planning by linear and convex models, *Optimization* **14** (1983), 513–35.

[241] Zsigmind, I., *Parametrization of all the Coefficients of a Linear Programming Problem*, Computing Center for Universities, ESZK4, Budapest, 1976.

[242] Zulehner, W., Schrittweitensteuerung fuer Einbettungsmethoden, Diss., Math. Inst., Universität Linz, 1981.

[243] Zulehner, W., A simple homotopy method for determining all isolated solutions to polynomial systems, *Math. Comput.* **50** (1988), 167–77.

[244] Zulehner, W., On the solutions to polynomial systems obtained by homotopy methods, *Numer. Math.* **54** (1988), 303–17.

[245] Jongen, H. Th. and Weber, G. W., *On Parametric Nonlinear Programming*, Preprint No. 5, March 1989, Lehrstuhl C für Mathematik, RWTH Aachen.

[246] Guddat, J., Jongen, H. Th., Kummer, B. and Nožička, F. (eds), *Parameteric Optimization and Related Topics II*, Akademie-Verlag, Berlin, to appear.

Glossary of Symbols and Some Assumptions

$C^k(\mathbb{R}^n, \mathbb{R}^m)$	space of k times continuously differentiable functions from $\mathbb{R}^n$ to $\mathbb{R}^m$
$C^k(U, V)$	(similar)
$Df(x)$	$\left(\dfrac{\partial f}{\partial x_1}, \ldots, \dfrac{\partial f}{\partial x_n}\right)$ Jacobian matrix $(x \in \mathbb{R}^n)$
$D^2 f(x)$	$\left(\dfrac{\partial^2 f}{\partial x_i \partial x_j}\right)_{i,j=1,\ldots,n}$ Hessian $(x \in \mathbb{R}^n)$
$D_x f, D_x^2 f$	partial Jacobian, Hessian
N	$N = \{0, 1, 2, \ldots\}$
det	determinant
$D^3 f(x)(v, v, v)$	$\displaystyle\sum_{i,j,k} \dfrac{\partial^3 f(x)}{\partial x_1 \partial x_j \partial x_k} v_i v_j v_k$
Ker B	$\{\xi = \mathbb{R}^m \,|\, B\xi = 0\}$, B a $n \times m$ matrix
$\mu \geqslant 0$	$\mu_j \geqslant 0$, all components
$\bar{A}$	closure of A
$\|x\|$	Euclidean norm, $\|x\|^2 = \displaystyle\sum_{i=1}^{n} x_i^2$
v^T	transposed vector
diag	diagonal matrix
$\perp$	orthogonal complement
H	$H := (h_1, \ldots, h_m)^\mathrm{T}$
G	$G := (g_1, \ldots, g_s)^\mathrm{T}$
$M[H, G]$	$M[H, G] := \{x \in \mathbb{R}^n \,|\, h_i(x) = 0, i = 1, \ldots, m; g_j(x) \leqslant 0, j = 1, \ldots, m\}$
LICQ	linear independence constraint qualification
MFCQ	Mangasarian–Fromovitz constraint qualification
LI, LCI	linear index, linear co-index
QI, QCI	quadratic index, quadratic co-index
$\mathscr{F}$	subset of $c^2(\mathbb{R}^n, \mathbb{R})$ consisting of non-degenerate functions

$\mathscr{F}(A)$ $\mathscr{F}(A) := \{f \in C^2(\mathbb{R}^n \times \mathbb{R}, \mathbb{R}) \mid \text{zero is a regular value for } (x, t) \to D_x^T f(x, t)\}$

$\mathscr{F}(B)$ $\mathscr{F}(B) := \{f \in C^3(\mathbb{R}^n \times \mathbb{R}, \mathbb{R}) \mid f \text{ satisfies condition B}\}$

$\mathscr{F}^*$ $\{(f, H, G) \in C^2(\mathbb{R}^n, \mathbb{R})^{1+m+s} \mid M[H, G] \text{ satisfies LICQ at all its points, and all critical points for } f|_{M[H.G]}\}$

$P(t)$ $P(t) := \min\{f(x, t) \mid h_i(x, t) = 0, i = 1, \ldots, m; g_j(x, t) \leqslant 0, j = 1, \ldots, r\}$

g.c. point   generalized critical point

Σ_{gc} $\Sigma_{\mathrm{gc}} := \{(x, t) \in \mathbb{R}^n \times \mathbb{R} \mid x \text{ is a g.c. point for } P(t)\}$

Σ_{gc}^i $\Sigma_{\mathrm{gc}}^i := \{(x, t) \in \Sigma_{\mathrm{gc}} \mid (x, t) \text{ of type } i\}, i \in \{1, \ldots, 5\}$

Σ_{stat} $\Sigma_{\mathrm{stat}} := \{(x, t) \in \mathbb{R}^n \times \mathbb{R} \mid x \text{ is a stationary point for } P(t)\}$

Σ_{stat}^i $\Sigma_{\mathrm{stat}}^i := \Sigma_{\mathrm{stat}} \cap \Sigma_{\mathrm{gc}}^i, i \in \{1, 2, 3\}$

Σ_{loc} $\Sigma_{\mathrm{loc}} := \{(x, t) \in \mathbb{R}^n \times \mathbb{R} \mid x \text{ is a local minimizer for } P(t)\}$

$\bar{\Sigma}_{\mathrm{loc}}^i$ $\bar{\Sigma}_{\mathrm{loc}}^i := \bar{\Sigma}_{\mathrm{loc}} \cap \Sigma_{\mathrm{gc}}^i, i \in \{1, \ldots, 5\}$

Σ_{reg} $\Sigma_{\mathrm{reg}} := \{(x, t) \in \Sigma_{\mathrm{gc}} \mid \mathrm{LI} = 0 \text{ at } (x, t)\}$

KKT system   Karush–Kuhn–Tucker system

KKT point   Karush–Kuhn–Tucker point

Σ_{KKT} $\Sigma_{\mathrm{KKT}} := \{(x, \lambda, \mu, t) \in \mathbb{R}^n \times \mathbb{R}^m \times \mathbb{R}^s \times \mathbb{R} \mid (x, \lambda, \mu) \text{ is a KKT point for } P(t)\}$

$\mathscr{F}^{**}$ $\mathscr{F}^{**} := \{(f, H, G) \mid \text{each point of } \Sigma_{\mathrm{gc}} \text{ belongs to type } 1, 2, 3, 4, 5\}$

Condition A   At every point of the closure of Σ_{stat} the MFCQ is satisfied

Condition B   Zero is a regular value for the mapping $\begin{pmatrix} x \\ t \end{pmatrix} \to \begin{pmatrix} D_x^T f(x, t) \\ \det(D_x^2 f(x, t)) \end{pmatrix}$

(E1)   There exists a continuous function $x : [t_A, t_B] \to \mathbb{R}^n$ such that $x(t)$ is a local minimizer for $P(t) (t_A < t_B)$

(E2) $x(0)$ is known

(V1)   There exists a neighbourhood U of $\{(x(t), t) \mid t \in [t_A, t_B]\} \subset \mathbb{R}^n \times [t_A, t_B]$ such that for all $(x, t) \in U$ the functions f, g_i and $h_j (i = 1, \ldots, m, j = 1, \ldots, s)$ are twice continuously differentiable with respect to x

(V2)   The LICQ is satisfied at $x(t)$ for each $t \in [t_A, t_B]$

(V3)   (The strong second-order sufficient condition) $D_x^2 L(z(t)) | T_{x(t)}^+ M(t)$ is positive definite for all $t \in [t_A, t_B]$ where $z := (x, t)$ (in particular $z(t) := (x(t), t)$),

$$T_{x(t)}^+ M(t) := \bigcap_{i \in I} \mathrm{Ker}\, D_x h_i(z(t)) \cap \bigcap_{j \in J^+(z(t))} \mathrm{Ker}\, D_x g_j(z(t)),$$

$$J^+(z(t)) := \{j \in J_0(z(t)) \mid \mu_j(t) > 0\}$$

and

$$L(z) := f(z) + \sum_{i \in I} \lambda_i h_i(z) + \sum_{j \in J} \mu_j g_j(z)$$

(A1) $(f, H, G) \in \mathscr{F}^{**}$

(A2) $(x^0, 0) \in \Sigma^1_{\mathrm{gc}}$ is known or easy to compute

(A3)   For all $t \in [0, 1]$, $M(t)$ is non-empty and there exists a compact set C containing $M(t)$

(B1)   A local minimizer for $P(0)$ is known.

(B2) $P(1) \equiv (P)$ where

$$(P) \qquad \min\{f(x) \,|\, h_i(x) = , i \in I, g_j(x) \leq 0, j \in J\}$$

(B3) $\psi(t) \neq \varnothing$ for all $t \in [0, 1]$ where $\psi(t)$ denotes the set of all global minimizers

(B4) $(f, H, G) \in \mathscr{F}^{**}$

(B4′)   Zero is a regular value of H

(B5)   For each t in some neighbourhood of zero, there is a unique solution $(x(t), \lambda(t), \mu(t), t)$ of $H(x, \lambda, \mu, t) = 0$

(B6)   MFCQ is satisfied for all $x \in M(t)$ for all $t \in [0, 1]$

(C1) $\varnothing \neq M := \{x \in \mathbb{R}^n \,|\, h_i(x) = 0, i \in I; g_j(x) \leq 0, j \in J\}$ and M is compact

(C2)   For each stationary point $\bar{x}$ of (P) it holds:

     (a) the LICQ is satisfied at $\bar{x}$,

     (b) $\mu_j > 0, j \in J_0(\bar{x})$,

     (c) $D^2 L(\bar{x}) | T_{\bar{x}} M$ is non-singular

(C3)   For each $r \in [r_1, r_2]$ and for each $x \in C(r)$ it holds:

     (a) the set $\{D_x h_i(x) \,|\, i \in I\}$ is linearly independent,

     (b) there exists a vector $\xi \in \mathbb{R}^n$ with

$$h_i(x) + D_x h_i(x)\xi = 0, i \in I,$$
$$g_j(x) + D_x g_j(x)\xi < 0, j \in J, \text{ with } g_j(x) \geq 0,$$
$$D_x g_{s+1}(x)\xi < 0 \text{ if } g_{s+1}(x) = r$$

(C4) $f, h_i, g_j, i \in I, j \in J$, are three times continuously differentiable

(D1) $F \in C^\infty(\mathbb{R}^n, \mathbb{R})$

(D2) F has compact level sets

(D3) x^0 is a global minimizer for $\min\{f(x) \,|\, x \in \mathbb{R}^n\}$ and x^0 is the only stationary point for $\min\{f(x) \,|\, x \in \mathbb{R}^n\}$, e.g. $f(x) := \|x - x^0\|^2$

$(\hat{P})$ $\min\{f(x) \,|\, g(x) \leq 0\}$ where $g(x)$ is defined by $g(x) := F(x) - F(\hat{x}) + \varepsilon$

$\hat{P}(t)$ $\min\{f(x, t) \,|\, g(x, t) \leq 0\}, t \in [0, 1]$ where

$$f(x, t) := f(x)$$
$$g(x, t) := g(x) + (t - 1)g(x^0)$$

(D4) $g(x^0) > 0$

(D5) $\hat{M} := \{x \in \mathbb{R}^n \,|\, g(x) \leq 0\} \neq \varnothing$

(D6) $(f, g) \in \mathscr{F}^{**}$

Index

active index set   2

characteristic numbers LI, LCI, QI,
    QCI   42
class $\mathscr{F}$ 23
class $\mathscr{F}(A)$ 25
class $\mathscr{F}(B)$ 27
class $\mathscr{F}^*$ 36
class $\mathscr{F}^{**}$ 42
C_s^k topology   23
complementarity condition   34
condition A   54
condition B   27
critical point   21, 23, 32
critical value   23

efficient point   14
Euler predictor   73

generalized critical point
    (g.c. point)   2, 31
g.c. point of type 1   42
g.c. point of type 2   43
g.c. point of type 3   44
g.c. point of type 4   46
g.c. point of type 5   48
goal realizer   167

JUMP I   132
JUMP II   140
jump at a point of type 2   118
jump at a point of type 3   122
jump at a point of type 4   125

Karush–Kuhn–Tucker point
    (KKT point)   1

Karush–Kuhn–Tucker system
    (KKT system)   1

linear independence constraint
    qualification (LICQ)   29
linear co-index (LCI)   34, 42
linear index (LI)   34, 42
local stability set   68, 93, 103
lower semicontinuous (l.s.c.) point-to-set
    map   148

Mangasarian–Fromovitz constraint
    qualification (MFCQ)   29
Morse lemma   21

non-degenerate critical point   21, 32, 34
non-degenerate function   22

PATH I   83
PATH II   96
PATH III   111
piecewise continuously differentiable
    (PC^1) map   36, 53
predictor   73
property efficient point with bound ε 14

quadratic co-index (QCI)   21, 34, 42
quadratic index (QI)   21, 34, 42

radius of convergence   59
realistic goal   167
regular point   23
regular value   24, 53

Sard's theorem   24

solution algorithm for one-parametric
   optimization problems 2
stationary index 40
stationary point 1, 32
strongly regular 61
strongly stable point 37

topological manifold

(with boundary) 30
transition point 70, 93, 103

upper semicontinuous (u.s.c.) point-to-set
   map 148

weakly efficient point, 15